AF541550

A TEXTBOOK OF PLANTATION CROPS

NIPA GENX ELECTRONIC RESOURCES & SOLUTIONS P. LTD.
New Delhi-110 034

About the Authors

Prof. Bhimasen Naik (b.1961) hails from an agri-horticultural family of Western Odisha (Village: Jamtalia, Tehsil: Sundargarh, District: Sundargarh). He obtained B.Sc. (Ag. & A.H.) and M.Sc. (Ag.) in Genetics and Plant Breeding from Chandra Shekhar Azad University of Agriculture and Technology, Kanpur in 1981 and 1983, respectively, and Ph.D. from Utkal University, Bhubaneswar in 1999. He was awarded ICAR Junior Research Fellowship and CSIR Senior Research Fellowship during his studies. He is a Fellow of the Indian Society of Genetics & Plant Breeding, New Delhi; and life member of the Crop Improvement Society of India, Ludhiana; the Indian Society of Plant Breeders, Coimbatore; the Society for Advancement of Rice Research, Hyderabad; the Indian Society of Oilseeds Research, Hyderabad and the Society for Plant Biochemistry & Biotechnology, New Delhi. He has worked on breeding and genetics of ginger, turmeric, mungbean, linseed and rice. He has more than 36 years of research and teaching experience at Odisha University of Agriculture and Technology. He has 56 research papers published in peer-reviewed national and international journals, 23 abstracts of symposia, 6 books, 1 book chapter (Springer-Nature) and 20 popular articles to his credit. He has guided one M.Sc. (Ag.) and one Ph.D. student. He is associated in the development of five cultivars in ginger (Suruchi), turmeric (Surama, Ranga and Rashmi) and linseed (Arpita); collaborated in development of three cultivars in rice (Ashutosh, Gobinda and Hasanta) and registered one genetic stock of mungbean (BSN 1, INGR 00011).

Dr. Ranjan Kumar Tarai (b.1977) obtained his B.Sc. (Ag.) in 1999 from College of Agriculture, Chiplima under Odisha University of Agriculture and Technology; M.Sc. (Hort.) in Fruits and Orchard Management and Ph.D. (Hort.) in Fruits and Orchard Management in 2002 and 2006 respectively from Bidhan Chandra Krishi Viswa Vidyalaya, Mohanpur, Nadia, West Bengal. He successfully qualified NET in Fruit Science during 2004 conducted by ASRB, New Delhi. He started his career in 2006 as Subject Matter Specialist (Horticulture) at Krishi Vigyan Kendra (Gajapati) and later on joined as Junior Scientist (Horticulture) at Regional Research and Technology Transfer Station, Bhawanipatna and as Programme Coordinator at KVK, Kalahandi under Odisha University of Agriculture and Technology, Bhubaneswar. Dr. Tarai is basically a Pomologist inclined to fruit research and teaching. He has an experience of over 14 years of teaching, research and extension. He has published more than 54 research papers in national and international journals, 7 Books, 26 popular articles, 18 book chapters, 20 extension booklets & Newsletters to his credit. He has guided four M.Sc. Ag. (Hort.) students. He is a life member of several societies in the field of Horticulture. He has devoted his works in development and transfer of technologies in fruits (especially minor fruit crops), vegetables and horticulture-based farming system suitable for resource poor area. Currently he is working as Associate Professor (Fruit Science) at College of Horticulture, Chiplima under Odisha University of Agriculture and Technology, Bhubaneswar.

Dr. Ajit Kumar Sahoo (b.1988) born at Jajpur district of Odisha and did his Graduation from College of Agriculture, Odisha University of Agriculture and Technology, Bhubaneswar in 2011, M. Sc. (Pomology) at Dr. Punjabrao Krishi Vidya Peeth, Akola in 2013 and Ph.D. (Pomology) from Odisha University of Agriculture and Technology, Bhubaneswar in 2017. He worked as Asst. Professor at College of Horticulture, Chiplima under Odisha University of Agriculture and Technology from 2015 to July 2020. Presently he is working as scientist at Coconut Research Station, AICRP on Palm, Bhubaneswar. He is actively involved in research on coconut as well as in teaching and extension activities. He has guided three M.Sc. Ag. (Hort.) students. He is working on micropropagation of pineapple, canopy management in sapota and value addition in banana. He has published 4 books, 2 book chapters, 2 popular articles, 19 research papers and 2 review papers in national and international journals/ scientific magazines.

Dr. Bijaya Kumar Sethy (b.1975) born in village: Bamara, Tehsil: Garadpur, District: Kendrapara. He obtained B.Sc. (Ag.), M.Sc. (Ag.) in Horticulture (Fruit Science) and Ph.D. from Odisha University of Agriculture and Technology, Bhubaneswar in 1996,1999 and 2021 respectively and successfully qualified National Eligibility Test (NET) in 2001 conducted by Agricultural Scientist Recruitment Board (ICAR), New Delhi. He started his career in 2006 as Subject Matter Specialist (Horticulture) at Krishi Vigyan Kendra (Nabarangpur) and later on joined as Assistant Professor (Horticulture) in 2010 at College of Agriculture, Chiplima under Odisha University of Agriculture and Technology, Bhubaneswar. He is basically a Horticulturist inclined to various horticultural research, teaching and extension. He has 16 years of research, teaching and extension experience and is presently working as Assistant Professor (Horticulture) at College of Agriculture, Chiplima under the Odisha University of Agriculture and Technology, Bhubaneswar, Odisha. He has published 12 research papers in various journals of national and international repute, 15 abstracts and 10 popular articles (Odia and English), 2 book chapters, 1 book and 3 booklets (Odia) and Newsletter to his credit. He is also life member of several societies in the field of horticulture and reviewer of different journals.

Dr. Sunil Samal (b.1974) obtained his B.Sc. (Ag.) from Shri Shahu Ji Maharaj University, Kanpur in 1997; M.Sc. (Ag.) in Horticulture from Odisha University of Agriculture and Technology, Bhubaneswar in 1999 and Ph.D. from Utkal University, Bhubaneswar in 2003; subsequently qualified NET in Fruit Science in 2004 conducted by ASRB, New Delhi. He started his career as Training Associate (Horticulture) at Krishi Vigyan Kendra, Dhenkanal in 2005 and later on joined as Programme Coordinator at Krishi Vigyan Kendra, Koraput. In 2010, he was selected for the post of Associate Professor (Post Harvest Management) and joined the College of Horticulture, Chiplima. In teaching he remained for eight years. Besides teaching he worked for the development of Horticulture, Horti-based farming system in the district. To his credit he accumulated an experience of over 16 years of Teaching, Research and Extension. He has published 14 research papers and 1 review paper in National and International journals, 10 popular articles, 6 Abstracts of Symposia, 1 book, 11 Extension Booklets and Newsletters to his credit. During his service carrier he attended a number of Refresher courses/Symposia/Seminars conducted by National and State level organizations. Presently, he is working as Senior Scientist (Horticulture) at the Regional Research and Technology Transfer Station, Dhenkanal.

A TEXTBOOK OF PLANTATION CROPS

Bhimasen Naik
Former Professor of Plant Breeding and Genetics
Odisha University of Agriculture and Technology
Chiplima Campus, Sambalpur-768 025, Odisha

Ranjan Kumar Tarai
Associate Professor of Fruit Science
Odisha University of Agriculture and Technology
Chiplima Campus, Sambalpur-768 025, Odisha

Ajit Kumar Sahoo
Assistant Professor of Horticulture
Odisha University of Agriculture and Technology
Bhubaneswar Campus, Bhubaneswar -751 003, Odisha

Bijaya Kumar Sethy
Assistant Professor of Horticulture
Odisha University of Agriculture and Technology
Chiplima Campus, Sambalpur-768 025, Odisha

Sunil Samal
Associate Professor of Horticulture
Odisha University of Agriculture and Technology
Mahisapat Campus, Dhenkanal-759 001, Odisha

NIPA GENX ELECTRONIC RESOURCES & SOLUTIONS P. LTD.
New Delhi-110 034

NIPA GENX ELECTRONIC RESOURCES & SOLUTIONS P. LTD.

101,103, Vikas Surya Plaza, CU Block
L.S.C.Market, Pitam Pura, New Delhi-110 034
Ph : +91 11 27341616, 27341717, 27341718
E-mail:newindiapublishingagency@gmail.com
www: www.nipabooks.com

For customer assistance, please contact
Phone: + 91-11-27 34 17 17 Fax: + 91-11- 27 34 16 16
E-Mail: feedbacks@nipabooks.com

ISBN: 978-93-91383-20-6

Composed and Designed by NIPA.

To

Abhipsa
Meerabai
Sri Baman Charan Sahoo
Biswajit
Shagun

Bhimasen
Ranjan
Ajit
Bijaya
Sunil

Preface

The Fifth Deans' Committee of Indian Council of Agricultural Research has recently revised the syllabus of B.Sc. (Hons.) Horticulture which is uniform throughout the country. The course 'Plantation Crops' is taught in the fourth semester. The present textbook covers the entire syllabus in 11 chapters. Simple and lucid language has been used for easy understanding of the beginners. The information contained in the book has been gathered from various published sources and internet websites which are mentioned at the end of each chapter under references. Attempts have been made to provide latest information; still some valuable information might have been missed. Questions are set at the end of each chapter to assess the understanding of the students.

We have tried our best to remove the errors, typographical or otherwise, from the text; still there might be some. We would highly appreciate if it is brought to our notice for rectification in next edition. We cherish the encouragement and cooperation received from our family members during preparation of the manuscript. We congratulate M/S NIPA, New Delhi for their support and publishing in a short time.

Though the book is primarily written for B.Sc. (Hons.) Horticulture students, the counterparts of B.Sc. (Hons.) Agriculture also may be benefitted. It may serve as a help book for post-graduate students.

We seek suggestions from learned teachers and students for further improvement of the book.

Authors

Contents

1

Plantation Crops An Overview

1.1 Introducton

Plantation crops which include perennial crops like coconut, arecanut, oil palm, cashew nut, tea, coffee, rubber, cocoa etc. have a unique role in the national economy of the country. Plantation crops are high-value crops of great economic importance and provide huge employment opportunity, especially to the women throughout the year. Plantation is the large-scale farm or estate in a tropical or semi-tropical country meant for farming that specializes in cash crops. The crops taken in large scale as plantation are termed as plantation crops. Most of the plantation crops are dollar earning crops. Apart from earning valuable foreign exchange, these crops form the backbone of the economy of the nation and the lifeline of many of the southern states, in particular, and provide basic raw materials for a number of industries. *The term plantation crop refers to those crops which are cultivated on an extensive scale in contiguous area, owned and managed by an individual or a company.* They are a group of commercial crops of perennial nature, cultivated extensively in tropical and subtropical situations which need employment of labour throughout the year and the produce of which are usually consumed after processing. They are high value commercial crops of greater economic importance and play a vital role in improving Indian economy, especially in view of their export potential, employment generation and poverty alleviation particularly in rural sector. Coconut, cashew nut, cocoa, areca nut, oil palm and palmyrah come under Ministry of Agriculture while tea, coffee and rubber are dealt by Ministry of Commerce. Plantation crops are highly income generating if managed properly. The cultivation of these crops was traditionally limited to corporate sector. The corporate revenue, however, did not percolate down to the benefit of the society. Plantations provide helps to our agriculture by way of high level of productivity and employment creation, apart from their catalysing contributions towards rural development. Being a labour-intensive enterprise, supplying modern technology and management tactics, plantation makes the optimum use of the marginal land resources. They generate considerable foreign exchange earnings by way of export.

1.2 Classification of Plantation Crops

1. Based on Botanical Relationship

1. **Monocot:** Family: Arecaceae, e.g., Coconut, Arecanut

2. **Dicot:**

Family	Examples
Theaceae	Tea
Rubiaceae	Coffee
Euphorbiaceae	Rubber
Sterculiaceae	Cocoa
Piperaceae	Black pepper
Lauraceae	Cinnamon
Anacardiaceae	Cashewnut

2. Based on Growth Pattern

Type	Examples
Vine	Betel vine, Black pepper
Shrub	Coffee, Tea
Tree	Rubber, Oil palm, Arecanut

3. Based on Extent of Growing

Grown as	Examples
Homestead plantation	Coconut, Arecanut, Black pepper
Estate plantation	Tea, Coffee, Rubber

4. Based on Intensity of Cultivation

Type	Examples
Single storeyed	Clove, Nutmeg
Multi-storeyed	Cashewnut, Arecanut, Turmeric, Ginger

1.3 Area and Production, Uses, History and Development

In case of coconut, the major plantation crop of national and international importance, the country is second in global production and first in productivity in terms of number of nuts produced per hectare of plantation. In case of areca nut, India is in an enviable position, being first in area, production and productivity. The fast rate of area expansion taking place in cocoa in recent years is indicative of the increasing acceptance and rapid growth of this sector. Plantation crops have a major role in the national agrarian economy, with annual contribution to the tune of Rs. 14,200 crores to the national GDP and foreign exchange earnings of about Rs. 2,440 crores. (Annual Report, CPCRI, 2018-19). The plantation sector has been considered to be a major *Source* of livelihood and employment for large section of the people (Joseph and George 2010). It is a highly labour-intensive sector with a high concentration of women workers (54 % in tea and coffee and 42 % in rubber) (Occupational wage survey 2006), comprises of labourers who have remained less developed, isolated, marginalized and vulnerable (Choudhary and Tayal 2010) and is a source of livelihood for small holders whose numbers are rising over the years.

Table 1.1: Percentage share of production of various horticulture crops

Crop	2015-16	2016-17
Fruits	31.5	31.5
Vegetables	59.1	59.3
Flowers & Aromatics	1.1	1.1
Plantation Crops	5.8	5.7
Spices	2.4	2.4
Total Horticulture	100	100

The plantation crops are confronted with natural calamities that resulted in considerable losses to the plantation sector. Plantation crops in India has a rich diversity and varied history with each crop having its own distinct historical and economic perspective of development; some of these crops are native to India and some others are originated from the new world tropics, which were subsequently introduced and naturalized in India. Despite the physical dissimilarities in the growth habits and requirements, and an incredible diversity in the nature of the economically useful parts of these plants, there are some constraints found concerning improvement in their plant type, environments in which they grow, their protection from biotic and abiotic stresses, and their post-harvest technology etc. During 1970, ICAR established the Central Plantation Crops Research Institute with headquarters at Kasaragod, by merging the Central Coconut Research Stations at Kasaragod and Kayamkulam and the Central Arecanut Research Station at Vittal along with its five substations at Kannara, Mohitnagar, Kahikuchi, Hirehalli and Palode. The Indian plantation sector has inherent strength of varied agro-climatic conditions, huge domestic demand, higher productivity, strong research and development network and technology dissemination systems. However, linkages between them for increasing the production and marketing efficiencies have not been fully utilized. Inclusive growth and sustainability of plantation economy could be achieved through integrated development of farming and industry coupled with a stable market. Plantation crops constitute an important segment of horticulture in Indian agriculture. Palms such as coconut (*Cocos nucifera* L.), oil palm (*Elaeis guineensis* Jacq.), and palmyrah (*Borassus flabellifer* L.) do contribute significantly to the rural economy of India.

Table 1.2: Crop wise area and production of plantation crops for three years

	Area in '000 Ha, Production in '000 MT					
Crop	2014-15		20115-16		2016-17	
	Area	Production	Area	Production	Area	Production
Plantation Crops						
Areca nut	450	747	474	714	466	730
Cashew nut	1030	745	1036	671	1035	779
Cocoa	78	16	81	17	83	19
Coconut	1976	14067	2088	15256	2092	15339
Total Plantation	3534	15575	3680	16658	3677	16867

Source: Horticultural Statistics at a Glance 2018 (www.agricoop.nic.in)

Table 1.3: All India area, production and productivity of plantation crops over the years 1991-92 to 2017-18

Year	Plantation Crops		
	Area ('000 ha)	Production ('000 MT)	Productivity (MT/ha)
1991-92	2298	7498	3.26
2001-02	2984	9697	3.25
2002-03	2984	9697	3.25
2003-04	3102	13161	4.24
2004-05	3147	9835	3.13
2005-06	3283	11263	3.43
2006-07	3207	12007	3.74
2007-08	3190	11300	3.54
2008-09	3217	11336	3.52
2009-10	3265	11928	3.65
2010-11	3306	12007	3.63
2011-12	3577	16359	4.57
2012-13	3641	16985	4.66
2013-14	3675	16301	4.44
2014-15	3534	15575	4.41
2015-16	3680	16658	4.53
2016-17	3677	16867	4.59
2017-18	3744	18082	4.83

Source: Horticultural Statistics at a Glance 2018 (www.agricoop.nic.in)

Table 1.4 Production share of leading plantation producing states for 2017-18

Sl No.	States / Union Territories	Production (in '000 MT)	% Share
1	Kerala	6054.79	33.48
2	Karnataka	436.08	27.30
3	Tamil Nadu	4234.92	23.42
4	Andhra Pradesh	1090.74	6.03
5	Maharashtra	361.07	2.00
6	Odisha	334.25	1.85
7	Assam	195.04	1.08
8	Gujarat	172.82	0.96
9	Bihar	52.88	0.29
10	Others	353.57	1.96
11	All India Total	18082.41	100.00

Source: Horticultural Statistics at a Glance 2018 (www.agricoop.nic.in)

The coconut palm which is referred as "Kalpavriksha" (Tree of heaven) as each and every part of the palm is used by human beings in one or the other ways. It provides food, drink, timber and fuel. Millions of people in India depend upon coconut for their livelihood. Coconut provides food security and livelihood opportunities to 20 million people around the globe and 10 million people in India (Kalpa, the CPCRI Newsletter, October- December,2017). Even though India is the largest coconut growing country in the world, its contribution to the export market is not significant. Coconut research became an integral part of the national agricultural research system of the country in 1966 when the Government of India abolished the Indian Central Coconut Committee and among the 'twin functionalities of Research & Development', the coconut research was taken over directly by the Indian Council of Agricultural Research and the development of the crop was entrusted to the Directorate of Coconut Development, which later transformed itself in to the Coconut Development Board in 1981. The potential list of coconut based value added products include desiccated coconut powder, coconut milk, coconut cream, coconut milk powder, coconut flour, coconut curd / yoghurt, coconut water concentrate, coconut jelly (Nata-de-coco), coconut vinegar, snow ball tender nut, virgin coconut oil, coconut chips, coconut pickle, coconut protein powder, coconut jam, coconut ice cream, neera and neera based value added products such as syrup, squash, honey, jaggery, candy and sugar. Although there is scope of coconut production potential of 27,300 nuts/ ha/year, the average coconut productivity in India is at a level of 10,345 nuts/ ha/year. This indicates the scope for increasing the productivity by bridging the yield gap, through supply of quality planting materials and adoption of better management practices.

Arecanut (*Areca catechu* L.) is one of the important commercial crops grown in parts of Karnataka, Kerala, Assam, Meghalaya, West Bengal, and Andaman and Nicobar Islands. The cultivation has also been extended to other states like Tamil Nadu, Andhra Pradesh and Maharashtra. It is popularly known as supari or betel nut, is grown for nuts which are used for chewing purpose. Areca nut is an important component of the religious, social and cultural celebrations and economic life of people in India. Areca nut is also used in ayurvedic and veterinary medicines. The habit of chewing areca nut is typical of the Indian subcontinent and its neighbourhood. India is the largest producer and consumer of areca nut in the world holding 62% of the area and 60% of the production. Areca nut being a profitable plantation crop, it is important to understand the package of practices and seasonal operations to be followed in the nursery, in young gardens and in old plantations to get maximum returns. To get additional income in the adverse periods, concept of cropping system is very important. More than four million people in different parts of the country, particularly in Karnataka, Kerala and Assam are dependent on arecanut industry for their livelihood.

Cashew (*Anacardium occidentale*) known as 'wonder nut', is one of the most valuable processed nuts traded on the global commodity markets and is also an important cash crop. The area, production and productivity of cashew have been increasing as a result of identification of superior clones, standardization of vegetative propagation techniques and near self-sufficiency in quality planting material. It has the potential to provide source of livelihood for the cashew growers, empower rural women in the processing sector, create employment opportunities and generate foreign exchange through exports. Cashew was introduced in India during the latter half of the sixteenth century for the purpose of afforestation and soil conservation. From its humble beginning as a crop intended to check soil erosion, cashew has emerged as a major foreign exchange earner next only to tea and coffee. Among various nuts such as hazelnuts, almonds, etc., cashew nut enjoys an unenviable position and it is an unavoidable snack in all important social functions especially in the western countries.

Oil palm is one of the highest oil yielding crops under equatorial bioclimatic conditions that could help the country to minimize the edible oil imports with the cultivation of location-specific oil palm hybrids along with the required management practices. The oil palm is the richest source of vegetable oil seed crops grown in the world. It yields 4-6 MT of oil per ha per annum. The oil palm comes to bearing in the third year and it yields two types of oils such as the palm oil obtained from mesocarp and the kernel oil from kernel. The economic life of oil palm is 25-30 years, though it lives up to 200 years. It is a crop of the future as source of health and nutrition, value addition, waste utilization, diversification, import substitution and sustainability. Realizing the potential of the crop, oil palm has been identified as the crop for solving the vegetable oil shortage in India. Under intensive management conditions providing recommended dose of fertilizer and weekly flood irrigation on good sandy loam soils, farmers may be able to realize an annual yield of around 20 tonnes of fresh fruit bunch (FFB), yielding about 4 tonnes of oil/ha.

Palmyrah palm is grown in the semi-arid regions of Tamil Nadu, Andhra Pradesh, Odisha, West Bengal, Bihar, Karnataka and Maharashtra. India has nearly 102 million palms and half of them are in Tamil Nadu. Palmyrah palm ranks first (among the available palm species) in yielding sugar as well as other edible and non-edible products. Many value-added products can be prepared from the fruit and other parts of the palm. There is a fairly good potential for the growth and development of the palm products-based industries in India. This dry land tree has been exploited from time immemorial for its nutritious neera and edible by-products. Every part is being used in day to day life. Amongst the edible palm products, jaggery, ripened palmyrah fruit, palmyrah tuber (starchy tuber which develops from the matured seed), palm candy and palm sugar and finally the fermented neera in the form of toddy are very important.

Tea (*Camellia sinensis*) belonging to family Theaceae and originated in South East China, is predominantly cultivated in Asia and Africa. It is the oldest known beverage. It fetches good foreign exchange and also provides gainful employment to a large number of people. In India tea is mainly grown in the states of Assam, West Bengal, Tamil Nadu and Kerala. India is the largest producer, consumer and exporter in Tea industry. The different tea species grown are: *C. assamica* (Assam tea), *C. sinensis* (China tea) and their natural hybrid *C. asamica* sub spp. *lasiocalyx* (Indo-China or Cambod type). Tea seed contains 20-45% oil rich in unsaturated fatty acid particularly linoleic acids and oil is used in cosmetic and pharmaceutical industries. The tea stimulates nerves and is good for health. High fluoride content of tea leaves prevents tooth decay. Tea has anti diabetic and anti-oxidant qualities. Green tea contains vitamin-K and vitamin B. Tea plantation sector has been considered to be a major source of livelihood and creation of employment opportunities in India.

Coffee is one of the most important beverages in the world prepared from roasted and ground beans. As an important plantation crop, it plays a very important role in the economy of many countries. More than 50 countries in the tropical and subtropical regions of Latin America, Africa, Asia involved in coffee production. There are two types of commercial coffee, *viz.*, Arabica coffee accounts 70 % and Robusta coffee accounts 30 % of world production.

It is estimated that coffee provides a livelihood for 25 million coffee farm families around the world. Caffeine is an important constituent of coffee which has stimulating effect. Arabica coffee is known for its aroma and low caffeine content. Robusta coffee is characterized by high caffeine content, is prepared for instant coffee.

Cocoa (*Theobroma cacao* L.) is the third important beverage crop grown for its seeds known as beans, used mainly in chocolate industry. It is the source of cocoa butter and cocoa mass. China and India are high potential consumption centres for huge population and also due to excellent economic growth rates in these countries. Cocoa industry in India has a demand gap of around 20,000 tonnes. Cocoa is grown as a companion crop in irrigated coconut and arecanut gardens in Kerala and Karnataka. Success of cocoa lies in low cost of production, comparative advantage of cocoa over competing crops and relative success in limiting the incidence of pests and diseases.

Commercial cultivation of natural rubber was introduced in India by the British. India, Thailand and Vietnam are among the important natural rubber producing countries in the world. Though India is one of the leading producers of rubber, it still imports rubber from other countries. Kerala contributes 90% of India's total production of natural rubber. Also, Kerala and Tamil Nadu together occupy 86% of the growing area of natural rubber. Rubber is used for a variety of purposes from erasing pencil marks to manufacturing of tyres, tubes and a large number of industrial products. Tapping is the process by which the latex is collected from a rubber tree.

The monoculture in small holdings is frightening during the year of slump and epidemic of pest and diseases. Raising more than one crops along with a particular plantation crop not only reduce the gestation period but also ensure steady and higher farm income even in the period of slump. Seasonal crops like vegetables in the initial years of the plantation crops and permanent crops like orange, arecanut, black pepper etc. could be grown in the matured plantation to augment productivity and profitability. Intercropping also stands as insurance against crop failure and price slump.

1.4 Constraints of Plantation Crop Cultivation and Processing

There are a number of constraints including lack of information and extension services, timely supply of processing facilities and lack of required knowledge of agronomic practices. Crop loss due to pests and diseases is a major production constraint and hence a crop protection umbrella for these major pests/pathogens would be crucial to increase production and productivity of these crops. High market price fluctuations in the case of coconut and its products are due to dependency on the price of one product – coconut oil - which is again dependent on the price of other vegetable oils. Thus, product diversification of coconut and development of value-added products become very important in the coconut industry. Inadequate training infrastructure to prepare the required skills among the farmers is also a reason for slow growth. Thus, the early stage of transition in the plantation economy requires effective planning strategies. The post-harvest management in plantation crops is very crucial. Perfect knowledge is essential in every stage of the production, *viz.*, production, harvesting, processing and marketing. Hence information technology becomes crucial. Capital requirement is particularly very high in the initial period. The common farmers find it difficult to invest the high amount without adequate institutional financial assistance, which is essential for the promotion of small-scale plantation crop cultivation.

1.5 Scope and Importance of Plantation Crops

1. Possible area expansion in traditional area is limited. So, there is great scope for expansion in non-traditional regions mainly in North and Eastern States where there is irrigation potential. Due to the development of drip irrigation facilities, new area and non-traditional area under plantation crops is increasing.
2. Changing plantation crops scenario warrants innovative strategies and approaches to address new challenges and march towards an accelerated growth of plantation sector to achieve competitiveness and sustainability. It becomes highly important for small holder plantation crops which contribute immensely as the mainstay of the agrarian economies of many states and Union Territories in the country.
3. Area occupied by plantation crops in India is very limited. Hence, there is good scope for its expansion.
4. There is a great scope of export earnings from plantation crops in India, e.g., coconut coir industry, tea, coffee and cashew industries.
5. Plantation crops can provide good employment opportunities. For example, more than 20 million people in rural areas are engaged in the production, processing and marketing the products of plantation crops like coconut, oil palm and palmyrah palm etc. Similarly, tea industry offers direct employment to 10 lakhs and indirect employment to 10 lakh people, while cashew processing factories alone provide employment to 3 lakhs people besides 2 lakhs farmers are employed in cashew cultivation.

6. They can provide good scope for diversification and thereby it can create sustainable agriculture.
7. Plantation industry supports many byproduct industries and also many rural industries. For example, coconut fibre (obtained from husk) industry.
8. Conservation of soil and ecosystem: e.g. tea and coffee with shade trees like silver oak planted on hill slopes and cashew plantation in barren and waste lands can protect soil from both water and wind erosion.

References

Choudhary N and Tayal D 2010. A comparative study of the informal condition of the plantation labourers of India and Sri Lanka. *Indian J. Labour Economics*, **53** (2): 339-357.

Horticultural Statistics at a Glance 2018. Government of India, Ministry of Agriculture & Farmers' Welfare, Department of Agriculture, Cooperation & Farmers' Welfare, Horticulture Statistics Division.

ICAR-CPCRI 2019. Annual Report 2018-19. ICAR-Central Plantation Crops Research Institute, Kasaragod- 671 124, Kerala, India, pp.164.

Joseph KJ and George PS 2010. *Structural Infirmities in India's Plantation Sector- Natural Rubber and Spices.* National Research Programme on Plantation Development. Centre for Development Studies, Thiruvananthapuram.

www.agricoop.nic.in

OUTCOMES ASSESSMENT

PART A

Answer the following questions true or false.

1. Plantation crops are highly income generating even if managed casually. True/False
2. Coffee is one of the most important beverages in the world. True/False
3. Commercial cultivation of natural rubber was introduced in India by the Portuguese. True/False
4. Tea is the oldest known beverage. True/False
5. Palmyrah palm is grown in the humid regions. True/False

PART B

Answer the following questions.

1. What are plantation crops?
2. Cocoa is the third/fourth important beverage crop.

3. The coconut palm is referred as "—".
4. There are two types of commercial coffee, *viz.*, — and —.
5. Why should more than one crops be raised along with a particular plantation crop?

PART C

Write a brief note on each of the following.

1. Classification of plantation crops.
2. Scope and importance of plantation crops.
3. Uses of plantation crops.
4. Constraints of plantation crop cultivation.
5. History and development of plantation crops.

2

Production Technology of Coconut

Botanical Name: *Cocos nucifera* L.
Family: Arecaceae / Palmae
Chromosome No: 32
Place of Origin: Malaysia
Common Name: Coconut

2.1 Introduction

Coconut (*Cocos nucifera* L.), is widely known as 'Kalpavriksha' (i.e., tree of heaven) as each and every part of the palm is useful in one way or the other and is an important perennial oil yielding crop of humid tropics. It provides food security and livelihood opportunities to 20 million people around the globe and 10 million people in India (Chowdappa *et al.*, 2017). Even though India is the largest coconut growing country in the world, its contribution to the export market is not significant. Coconut as an edible oil and food crop, is now slowly transforming into raw material for various pharmaceutical, nutraceutical and cosmetic products. Coconut is rich in fibre, vitamins and minerals. Because of its health benefits beyond its nutritional content, it has been considered as "functional food". Hence, there exists a huge scope for coconut-based agri-business in India, which will increase the present 8% level of value addition to 25%. The coconut-based value-added products are desiccated coconut powder, coconut milk, coconut cream, coconut milk powder, coconut flour, coconut water concentrate, coconut jelly (Nata-de-coco), coconut vinegar, virgin coconut oil, coconut chips, coconut pickle, coconut ice cream, neera and neera based value added products such as syrup, squash, honey, jaggery, candy and sugar (Chowdappa *et al.*, 2017). India ranks third in area and first in production of coconut in the world. During 2016-17, the annual coconut production in India is 23.90 billion nuts from an area of 2.08 million ha with an average productivity of 11,481 nuts/ha (Subramanian *et al.*, 2018). Philippines, Indonesia, Ivory Coast, Malaysia, Thailand, Oceania, Papua New Guinea and Sri Lanka are the other leading coconut growing countries. In India, more than 90 % of the coconut acreage and production lies in the four southern states, namely, Kerala, Tamil Nadu, Karnataka and Andhra Pradesh. The unopened spathe is tapped for toddy; fresh toddy called "Neera" is a tonic. Whereas sweet toddy can be converted into jaggery and sugar, fermented toddy is a mild alcoholic drink and vinegar can also be prepared from it. Water from tender coconut is a refreshing and delicious drink.

2.2 Climate and Soil

The coconut palm is grown under varying climatic and soil conditions. It is essentially a tropical plant, growing mostly between 20° N and 20° S latitudes. The ideal mean temperature for coconut growth and yield is 27±5°C and relative humidity more than 60%. Very high humid conditions during growth of palms are not considered ideal. The coconut palm grows well up to an elevation of 600 m above mean sea level. A well distributed rainfall of about 200 cm per year is best for proper growth and high yield. In areas of insufficient rainfall and irregular distribution, irrigation is desirable. The palm requires plenty of sunlight and it does not grow well under shade or in too cloudy regions. About 2000 hrs of sunshine in a year is considered necessary for the healthy growth of the palm.

Coconut is grown in different types of soil such as laterite, coastal sandy, alluvial and also in reclaimed soils of the marshy lowlands. It tolerates salinity and a pH range from 5.0-8.0. Soil with a minimum depth of 1.2 m and fairly good water holding capacity is preferred for coconut cultivation. The ideal coconut growing soils are well drained and aerated soil with a minimum depth of 80 to 100 cm, with adequate nutrient availability and water holding capacity.

2.3 Choice of Cultivars

Coconut cultivars are classified into two groups, i.e., Tall and Dwarf.

Tall cultivars: Tall type cultivars grow to a height of 15 to 18 m and they can live from 50 to 80 years. They produce copra of high quality and have higher quantity of oil as compared to dwarf cultivars. The commonly grown tall cultivars in India are East Coast Tall, West Coast Tall, Tiptur Tall etc.

The important tall cultivars of coconut are mentioned below:

Sl No.	Cultivars	Yield (nuts/ palm/year)	Suitable region for cultivation
1	West Coast Tall	80	Kerala, Karnataka, Gujarat, Bihar, Madhya Pradesh, Lakshadweep, Odisha, Tamil Nadu and Tripura
2	East Coast Tall	70	Tamil Nadu, Andhra Pradesh, Bihar, Odisha, West Bengal, Andaman and Pondicherry
3	Tiptur Tall	86	Karnataka
4	Chandrakalpa (Laccadive Ordinary)	100	All States
5	VPM-3 (Andaman Ordinary)	94	Andaman, Andhra Pradesh, Assam, Bihar, Odisha, Madhya Pradesh, Kerala, Pondicherry, Tamil Nadu, Tripura and West Bengal
6	Benaulim Green Round	150	Maharashtra and Goa
7	Kamrupa (Assam Green Tall)	101	Assam and North Eastern States
8	Kerachandra (Phillippines Ordinary)	110	All States
9	Kerasagar (Seychelles)	100	Kerala
10	Aliyar Nagar-1(ALR-1) (Arasampatti Tall)	126	Tamil Nadu and Pondicherry

Source: Dhanpal and Thamban (2007)

Dwarf cultivars: Dwarf cultivars of coconut are shorter in height (5-7 m) and life span (40-50 years) as compared to tall cultivars. They come into bearing stage after 3 to 4 years of planting producing nuts of smaller size and the copra has low oil content. The commonly grown dwarf cultivars of coconut in India are: Chowghat Green Dwarf (CGD), Chowghat Orange Dwarf (COG), Chowghat Yellow Dwarf (CYD), Gangabondam (GBGD), Malayan Green Dwarf (MGD), Malayan Orange Dwarf (MOD), Malayan Yellow Dwarf (MYD). Among these, Chowghat Orange Dwarf (COD) has very good quality of tender nut water and has been released as tender nut cultivar suitable for cultivation in all states.

Hybrids: Hybrids are intervarietal crosses of two morphological forms of coconut. They show earliness in flowering and give high yield, better quality copra and oil than their parents. When the Tall cultivars are used as female and dwarfs as male, the hybrids are known as T×D and the reciprocal is known as D×T hybrids.

Sl No.	Hybrid	States where recommended	Average Yield (nuts/ palm/year)	Released by
1	Kera Sankara (WCT×COD)	Kerala, Karnataka, Coastal Maharashtra, Coastal Andhra Pradesh	106	CPCRI
2	Chandra Sankara (COD×WCT)	Kerala, Karnataka, Tamil Nadu	116	CPCRI
3	Chandra Laksha (LCT×COD)	Kerala, Karnataka	109	CPCRI
4	Kera Ganga (WCT×GBGD)	Kerala	100	KAU
5	Laksha Ganga (LCT×GBGD)	Kerala	108	KAU
6	Ananda Ganga (ADOT×GBGD)	Kerala	95	KAU
7	Kera Sree (WCT×MYD)	Kerala	140	KAU
8	Kera Sowbhagya (WCT×SSAT)	Kerala	130	KAU
9	VHC-1 (ECT×MGD)	Tamil Nadu	98	TNAU
10	VHC-2 (ECT×MYD)	Tamil Nadu	107	TNAU
11	VHC-3 (ECT×MOD)	Tamil Nadu	156	TNAU
12	Godavari Ganga (ECT×GBGD)	Andhra Pradesh	140	APAU
13	Kalpa Samrudhi (MYD×WCT)	Kerala, Assam- Drought tolerant	117	CPCRI
14	Kalpa Sankara (CGD×WCT)	Root (wilt) disease prevalent tracts	85	CPCRI
15	Kalpa Sreshta (MYD×TPT)	Kerala and Karnataka	167	CPCRI

Source: Thomas *et al*., (2010) and Dhanpal and Thamban (2007)

2.4 Propagation

Coconut is propagated by seed nuts. Selection of seed nuts and seedlings is of utmost importance in coconut as the performance of the new progeny can be evaluated only several years after planting.

2.4.1 Mother Palm Selection

For raising new coconut plantations, seed nuts are to be collected from selected mother palms. Mother palms should be of 20 years old or more. It is advisable to select middle-aged palms as it is easier to identify good yielder from poor yielder.

The important criteria for selection of mother palm are:

a) Trunk should be straight and stout with even growth and closely spaced leaf scars,

b) Spherical or semi-spherical crown with short fronds (with production of more than 30 leaves and 12 inflorescences carried evenly on the crown),

c) Short and stout bunch stalks without tendency to drooping,

e) Inflorescence should be having 25 or more female flowers,

f) Constant yield of about 80 nuts under rainfed conditions and 120 nuts under irrigated conditions, 150 g copra per nut and

h) Tolerance to major pest and diseases.

2.4.2 Collection of Seed Nuts and Sowing

Seed nuts are collected during April-May and sown during June in west coast region of India, whereas in the east coast region, nuts are sown during October-November. However, when irrigation facilities are available seed nuts can be collected and sown at any period of time depending on the requirement. Only fully matured nuts of about 12 months old should be harvested. Care should be taken not to damage the seed nuts while harvesting. Too big or too small nuts and nuts of irregular shape and size should be discarded in the bunch. Seed nuts of tall cultivars are sown one or two months after collection and those of dwarf cultivars are sown either immediately or within 10-15 days after harvest.

2.4.3 Raising Nursery

Nursery site should be well-drained with coarse textured soil near reliable irrigation water source. The seed nuts can be sown in flat beds if there is no drainage problem. The seeds are to be sown in raised beds if water stagnation is a problem. Nursery can be raised either in the open area with artificial shade or in gardens where the palms are tall and the ground is not completely shaded. The seed nuts should be planted in long and narrow beds at a spacing of 40 × 30 cm during May-June or September-October, either vertically or horizontally in 20-25 cm deep trenches. Advantage in vertical planting is less damage during transportation. However, in delayed planting when the nut water goes down considerably it is good to go for horizontal sowing for better germination.

2.4.4 Selection of Seedlings

Seed nuts, which do not germinate within 5 months after sowing as well as those with dead sprouts should be removed. Only good quality seedlings (9-

12 months) should be selected by a rigorous selection based on the following characteristics:

1. Early germination, rapid growth and seedling vigour.
2. Six to eight leaves for 10–12-month-old seedlings.
3. Collar girth of 10 cm and above.
4. Early splitting of leaves.

2.4.5 Poly Bag Nursery

Good quality seedlings can be raised using poly bags. Transplant germinated seed nuts in poly bags (500-gauge thickness) of 45 × 60 cm with 8-10 holes at the bottom. The commonly recommended potting media are top soil mixed with sand (3: 1) or top soil, sand or coir dust and well rotten and powdered farm yard manure (3: 1: 1). Potting mixture containing sand + vermicompost (3: 1) is also ideal for raising poly bag seedlings. As the entire ball of earth with the root system can be placed in the pits, there will not be any transplanting shock for the seedlings.

2.5 Land Preparation and Planting

The soil of the site should have a minimum depth of 1.2 m with good water holding capacity. Shallow soils with underlying hard rock, water stagnation should be avoided. The site should have adequate well distributed rain or dependable water source. In laterite soil with rocky substratum, pits of size 1.2 × 1.2 × 1.2 m are to be dug and filled up with fertile top soil, powdered farm yard manure and ash up to a depth of 60 cm before planting. Addition of 2 kg of common salt will help in loosening the soil in such areas. Two layers of coconut husk are arranged with concave surface facing up at the bottom of the pit before filling up the soil. This will help in conserving soil moisture. At the centre of the pit, remove the soil mixture and plant the seedling. The soil well around the seedling is pressed and the seedlings are provided with shade by using coconut leaves or palmyrah leaves. A spacing of 7.5 × 7.5 m is generally recommended for tall cultivars of coconut. If triangular system is adopted, an additional 20 to 25 palms can be planted. Planting the seedlings during May-June, with the onset of pre-monsoon rains is ideal. Under assured irrigation, planting can be done during April also. In low-lying areas, plant the seedlings in September after the cessation of heavy rains. Planting can also be taken up in Tamil Nadu with the onset of north-east monsoon.

Sl No.	Spacing (m x m)	System of planting	No. of plants
1	7.5 × 7.5	Square	175
2	7.5	Triangular	205
3	9 × 6.5	Rectangular	170

Planting systems

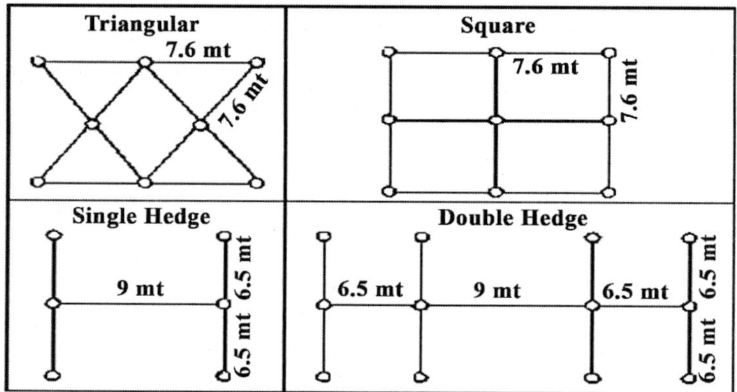

2.6 Nutrition

Coconut flowers and fruits throughout the year. Therefore, its requirement of water and nutrients should be supplied throughout the year. Nutrient exhaust from one hectare of coconut ranged from 92 to 149 kg N, 12 to 20 kg P and 119 to 183 kg K. This clearly indicates that K and N are required in higher quantities for coconut production. Integrated nutrient management includes the intelligent use of organic, inorganic and online biological resource (BNF) so as to sustain optimum yields, improve or maintain soil's chemical and physical properties and provide crop nutrition packages which are technically sound, economically attractive, practically feasible and environmentally safe (Tandon, 1990). Nutritional management practices based on soil test reports would help to overcome non-judicious application of fertilizers and in turn would improve soil health, crop health and the yield.

Regular manuring is essential for better establishment of the palm.

Table 2.1: Recommended fertilizer application in coconut

Age	May-June			September-October		
	N	P_2O_5	K_2O	N	P_2O_5	K_2O
1st year	-	-	-	50	40	135
2nd Year	50	40	135	110	80	270
3rd Year	110	80	270	220	160	540
4th Year onwards	170	120	400	330	200	800

Source: CPCRI, Kasaragod

As per CPCRI recommendation, a manurial and fertilizer dose of 500 g N (1 kg Urea), 320 to 330 g P_2O_5 (1.5 kg Rock Phosphate or 2 kg super phosphate) and 1200 g K_2O (2 kg Muriate of Potash) along with 50 kg of farm yard manure and 3 to 5 kg neem cake is required by a coconut palm annually. One-third of the dose of fertilizers is applied at the beginning of South West Monsoon and the rest two-thirds are applied after cessation of rain, i.e., during September-October. The first dose of fertilizer is applied after 3 months of planting and the dose is gradually enhanced and full dose is applied from 4th year onwards. The fertilizers are applied around the palm within the radius of 1.8 m and forked in. A trench (25 cm depth) around the circular basins is dug at 1.8 m radius and then fertilizers are applied and covered with soil. Soon after fertilizer application, the plants are irrigated. Green manure crops like *Pueraria*, cowpea, sunhemp, *Calopogonium* or dhaincha may be sown and ploughed *in situ* at the time of flowering as a substitute of compost to be applied. The green manure crops are sown immediately after the receipt of monsoon rains (May-June and September-October by South-West monsoon and North-East monsoon, respectively) according to the soil and climatic conditions @ 50 g/palm in the basin and incorporated before flowering.

Integrated nutrient management by using 2/3rd recommended fertilizer dose along with recycling of biomass by vermicomposting gives the best economic benefit in a sustainable manner (Palaniswami *et al*., 2007).

2.7 Water Management

Irrigation is required by the coconut palm from November to May. The method of irrigation followed in coconut is flood, basin and sprinkler etc. Young palms are irrigated with 45 litres of water and 200 litres for adult palms once in 4 days. Drip irrigation is beneficial in water scarce areas through which 32 litres of water per day are sufficient. Under northern Kerala conditions, 32 litres/palm/day is sufficient (66 % of the evaporation requirement) to achieve higher yield. Another advantage of drip irrigation is that fertilizers can also be applied along with irrigation water (fertigation). To supply 32 litres/day, four drippers at a discharge rate of 4 litres/hour will be required, so that daily two hours of irrigation is sufficient. In water logged area, provide drainage by digging wide drains between the rows of palms and by raising the level of ground around the individual palms. Mulching with composted coir pith to 10 cm thickness (approximately 50 kg/palm) around coconut basin is also ideal method to conserve moisture.

2.8 Weed Management

Generally, weeding is carried out when the monsoon recedes preferably during September-October. Weeding can be done either by hand or with a sickle depending on the intensity of weed growth. The weeds can be removed by hand around the palm base and the weeds in the interspace need only be slashed with sickle. Clean weeding may be avoided. While weeding, dried shoots and other thrashed materials can be used as mulch around the base of palms, which will help to conserve moisture in the ensuing dry months and help in vermicomposting process in the basin as well as in the interspaces.

2.9 Coconut Based Cropping System

Coconut based cropping systems involving annuals, perennials and combinations of both annuals and perennials have been developed to suit the availability of resource like labour, rainfall and irrigation facilities, finance, soil characteristics and the farmers' needs and market demands. The crops like cowpea, pumpkin, ash gourd, amaranthus, okra, turmeric, ginger, black pepper, pineapple etc. are found to be suitable intercrops which can be grown organically in the interspaces of coconut garden. Mixed farming system integrating crop husbandry and livestock is an integral part of the organic production system. It involves cultivation of shade tolerant fodder crops in the interspaces of coconut and integrating animal enterprises like dairy, poultry, fisheries etc. and recycling the by-products. Care should be taken for maintenance of livestock and other components as well as production of fodder (feed) needed for them are based on organic standards. Such a system enables the use of by-products in the farm itself without depending on external source.

High density multispecies cropping system (HDMSCS), is one of the mixed cropping systems, where a number of compatible crops are grown in a unit area to meet the diverse needs of the farmers such as food, fuel, timber, fodder and cash and are ideally suited for smaller units of land and at maximum production both temporally and spatially. This also leads to control of weeds, soil and water conservation, better microclimate and favourable microbial activity in the soil.

Cropping sequence	Place of Work	Achievement out of the system	Reference
Coconut + Cocoa + Banana + Moringa + Pineapple	AICRP, Aliyarnagar	This cropping system with 75 % NPK + organic recycling with vermicompost recorded highest nut yield of 182 per palm and highest net income (Rs. 3.80 lakhs per ha) and B:C ratio (2.71)	Nimbolkar *et al.* (2016)
Coconut + cocoa + lime + banana + drumstick	AICRP, Arasikere	With all physical and chemical quality of coconut, this system recorded net income of Rs. 2, 94,810 per hectare compared to mono crop (Rs. 68,200/ha).	Roy *et al.* (2001)
In an area of 1 ha, 150 coconut (7.5 × 7.5 m) + black pepper (1.25 m away from coconut base–150 vines) + cocoa (2.5 m between 2 rows of coconut – 525 plants) + pineapple (1-2 m in the rows, two rows of pineapple-4900 plants)	Coastal region of southern state of India	This model recorded higher yield of coconut (20%) and net returns compared to mono cropping of coconut, besides enhancing soil fertility due to recycling of by-products.	Khan and Krishnakumar (2002)

2.10 Crop Protection

Being a perennial crop, coconut is attacked by a number of pests and diseases throughout the year. However, only a few are considered to be of economic importance. All parts of the palm, *viz*., leaves, stem, root, inflorescence and nuts are subjected to attack by pests. The major insect pests of coconut are rhinoceros beetle (*Oryctes rhinoceros*), red palm weevil (*Rhynchophorus ferrugineus*), leaf eating caterpillar (*Opisina arenosella*), root eating white grub (*Leucopholis coneophora*), coried bug (*Paradasynus rostratus*) and coconut eriophyid mite (*Aceria guerreronis*).

2.10.1 Insect-Pests

The pests and their management in coconut as described by Subramanian *et al.* (2018) is described as follows:

1. Rhinoceros beetle *(Oryctes rhinoceros)*

The adult beetle bores through into the unopened fronds and spathes. The affected frond shows the characteristic diamond shaped cuts (V-shaped) upon unfurling as well as exposing chewed up fibres from the feeding site. Infestation on spathes destroys the inflorescence and thus prevents production of nuts. It breeds in a variety of materials such as decaying organic debris, farmyard manure, compost pit etc. It may also bore into the collar region of the young palms resulting in dead heart, twisted spindle with elephant-tusk like symptoms and distorted leaflets. In adult palms, the beetle bores into the unopened fronds/spindle region/spear leaf/spathes/nuts.

Management

1. Field sanitation by removal of decaying organic debris.
2. Mechanical extraction of beetles with hooks without causing any injury to the growing point of the palm.
3. Application of powdered neem cake @ 250 g mixed with equal quantity of sand in top most three leaf axils three times a year or fill the inner most two leaf axils with 12 g naphthalene balls covered with sand at 45 days interval.
4. Placement of three perforated sachets containing chlorantraniliprole a.i. 0.4% (5 g) or fipronil (3 g) or three botanical cake (2 g each) developed by ICAR-CPCRI during monsoon phase. During dry period, 100 ml of water may be poured over the sachet for effective release of the molecule.
5. The breeding sites may be treated with green muscardine fungus (*Metarhizium anisopliae).* The fungus can be mass multiplied on local materials such as coconut water, cassava chips and semi-cooked rice.
6. Application of mixture of either neem seed powder + sand (1 : 2) @150 g/palm or neem seed kernel powder + sand (1 : 2) @150 g per palm in the base of the three inner most leaves in the crown.

2. Red palm weevil *(Rhynchophorus ferrugineus)*

Young palms (below 20 years) are susceptible to damage by this pest. Bud rot and leaf rot disease and rhinoceros beetle attack are predisposing factors for red palm weevil infestation. Wilting of the central spindle, presence of chewed fibres and cocoons in the trunk, presence of holes in the trunk with brown fluid oozing out are the important symptoms.

Management

1. Avoid injury to the palms, as it would attract the weevil to lay eggs. Mechanical injury caused, if any, should be treated with coal tar. While cutting of fronds, leave petiole to a length of 120 cm from the trunk to prevent the entry of weevils through the cut end.
2. Filling of the inner most two leaf axils with 12 g naphthalene balls covered with sand at 45 days interval.
3. Coconut petiole pieces smeared with fermented toddy kept in pots @ 10 pots/ ha also serve as weevil traps. The traps should be placed at evening and the weevils trapped should be destroyed in the next morning.
4. Install pheromone traps @ 1 trap/ha on coconut trunk at a height of 2 m from ground to mass trap and destroy the weevils. This technology should be taken up on community basis.

3. Leaf eating caterpillar *(Opisinia arenosella)*

The caterpillars live on the under surface of leaflets and feed voraciously on the chlorophyll containing functional tissues. This affects the health of the palm adversely by reducing the photosynthetic area and results in reduction of yield. The severity of infestation by this pest is more during the summer months from February to June. With the onset of South West monsoon, the pest population of the leaf eating caterpillar infested coconut garden begins to decline. In severe outbreaks of leaf eating caterpillar, the older leaves of the palms are reduced to dead brown tissue and only three or four youngest leaves at the centre of the crown remain green.

Management

1. Heavily affected and dried outer most 2 to 3 leaves should be cut and burnt.
2. Periodical release of larval parasitoids such as *Goniozus nephantidis* @ 20 parasitoids /palm, *Bracon brevicornis* @ 30 parasitoids/palm.
3. The release of pre-pupal parasitioid *Elasmus nephantidis* @ 49 per cent and pupal parasitioid *Brachymeria nosatoi* @ 32 per cent, respectively, for every 100 pre-pupae, pupae estimated to be present on the palm.
4. Badly infested leaves are cut and burnt and then the palms should be sprayed with 0.02% dichlorvos @ 2 ml in 10 litres of water.

4. Coried bug *(Paradasynus rostratus)*

The adults and nymphs feed by desapping the contents on buttons and developing nuts below the perianth region. The feeding points develop to brownish necrotic lesions, which later turn to furrows or cracks. The symptoms are easily identified by cracks and gummosis. Severe damage leads to nut fall and malformation of mature nuts.

Management

1. Spraying of azadirachtin @ 0.004% (4 ml/L) reduces the pest incidence. Two rounds of azadirachtin spray on 1-5 months old coconut bunches during May-June and September-October may control of the pest satisfactorily.
2. The red ant, *Oceophylla smaradigna* has been found to have antagonistic effect on the pest.
3. Spraying Lambda-cyhalothrin @ 1ml/litre on the pollinated bunches in case of severe outbreak.

5. Coconut eriophyid mite *(Aceria guerreronis)*

The mites infest by sucking sap from the soft meristematic tissues underneath the tepals on buttons. In the initial stages, symptoms are seen as triangular patches close to perianth. Later because of the continuous sucking of sap by various stages of mites present beneath the inner bracts of perianth, brown coloured patches are formed. As the nuts grow in size, the injured patches become warts and then develop into longitudinal splits on the surface of nuts. The damage thus caused affects the quality of husk and de-husking becomes difficult.

Management

1. Removal of dried spathes, inflorescence parts, fallen nuts etc. and burying them in the soil or burning them reduces the pest inoculum and consequent infestation. Crown cleaning should be taken up as and when necessary.
2. Root feeding azadirachtin 10,000 ppm @ 10 ml + 10 ml water.
3. Spraying with neem oil-garlic-soap mixture @ 2 per cent concentration (neem oil 200 ml, soap 50 g and garlic 200 g mixed in 10 litres of water) is effective.

4. Application of 5 kg neem cake along with the recommended dose of chemical fertilizers and 50 kg cow dung or compost, per palm should be applied. Three sprayings of palm oil (200 ml) and sulphur (5 g) emulsion on the terminal five pollinated coconut bunches during January-February, April-May and October-November reduces mite incidence drastically.
5. Sulphur is recommended for spraying at 0.4 percent concentration for mite control (5 g per litre of water). Spraying of either 0.2 percent triazophos / 0.1 percent dicofol has also been found to be effective management of the mite based on field trials.
6. Kalpa Haritha (a tall selection from Kulasekaram green dwarf) recorded lowest mite incidence in the field and could be a preferred choice in endemic zones.

6. Rugose spiralling whitefly *(Aleurodicus rugioperculatus)*

It was found to feed and breed profusely from the under surface of the palm leaves. Coconut and banana are its most favoured host plants. It could not cause greater economic damage, however, it produced enormous quantities of honeydew on coconut and other intercrops in the palm system, resulting in heavy deposition of sooty mould (*Leptoxyphium* sp.) on affected plants. On coconut, sooty mould deposition was confined to the upper surface of coconut leaves including midrib impairing photosynthesis. Presence of black sooty mould is one of the characteristic symptoms of feeding damage by this pest and that also ascertained the presence of it in the palm system.

Management

1. Application of 1% starch solution on leaflets to crumble out the sooty moulds.
2. In severe case, spray neem oil @ 0.5% and no insecticide is recommended.
3. Installation of yellow sticky traps on the palm trunk to trap adult whiteflies.
4. Buildup of the *Aphelinid* parasitoid (*Encarsia guadeloupae*) and reintroduction of parasitized pupae to emerging zones of whitefly outbreak.
5. Conserve the natural population of the green lacewing, *Pseudomallada* sp. and other lady beetles to suppress the pest population.
6. *In situ* habitat conservation of the sooty mould scavenger *Leiochrinid* beetle (*Leiochrinus nilgirianus)* and introductory bio-scavenging/ biological control in new areas of its emergence.

2.10.2 Diseases

The coconut palm is affected by a number of diseases, some of which are lethal while others gradually reduce the vigour of palms causing severe loss in yield.

1. Bud rot

It is caused by *Phytophthora palmivora*. The earliest symptom is the yellowing of one or two younger leaves surrounding the spindle. The spindle withers and droops down. The affected spindle leaf turns brown, hangs down and can easily be pulled out as the basal portion of the spindle is completely rotten emitting a foul smell. Subsequently younger leaves next to the spindle also fall away one by one leaving only outer whorl of matured leaves in the crown. The bud rot pathogen also causes water-soaked lesions on nuts, quite independent of bud rot, and nut fall commonly called as 'Mahali'. The disease is more prevalent during monsoon when the temperature is low and humidity is high.

Management

1. All disease affected tissues should be burnt and removed from the palm.
2. Cleaning of crown and Bordeaux mixture (1%) application to palms before the onset of monsoon preferably in the last week of May or first week of June. About 300 ml Bordeaux mixture (1%) or chlorothalonil solution (3 g in 300 ml water) may be poured in the base of the spindle.
3. Rhinoceros beetle attack predisposes the palm to bud rot infection. Hence prophylactic measures to avoid beetle infestation have to be undertaken in bud rot endemic area.

2. Root (wilt) disease

It is caused by *Phytoplasma* and is transmitted by lace bug (*Stephanitis typica)* and the plant hopper (*Proutista moesta)*. The important visual diagnostic symptoms of the disease are abnormal bending or ribbing of the leaflets termed as 'flaccidity', a general yellowing and marginal necrosis of the leaflet and unopened inflorescence. The nuts are smaller and the kernel is thin. The oil content of copra is also reduced.

Management

1. Application of organic manure (25 kg farm yard manure or 10 kg vermicompost), biomass recycling by application of leguminous green manure crops and glyricidia leaves, soil test-based application of fertilizers (500 g N, 300 g P_2O_5, 1250 g K_2O and 1 kg $MgSO_4$ / palm

/ year) in two splits, soil moisture conservation, adequate drainage, intercropping and mixed farming coupled with recycling of organic matter.

2. Adoption of recommended management strategies for leaf rot disease, rhinoceros beetle and red palm weevil.
3. Growing of tolerant cultivars like Kalparaksha (selection from Malayan Green Dwarf), Kalpasree (selection from Chowghat Green Dwarf) and the hybrid Kalpa Sankara (Chowghat Green Dwarf × West Coast Tall) released from ICAR- CPCRI are suitable for cultivation in root (wilt) disease endemic tracts.

3. Stem bleeding disease

Stem bleeding is caused by fungus, *Thielaviopsis paradoxa.* Imbalanced nutrition, excess salinity etc. are the predisposing factors. The disease is characterized by the exudation of dark reddish-brown liquid from the longitudinal cracks in the bark generally at the base of the trunk. The bleeding patches spread throughout as the disease advances. The liquid oozing out dries up and turns black. Leaves in the outer whorl turn yellow, droop and dry. Nut fall also is seen.

Management

1. Apply a paste of talc-based formulation of *Trichoderma harzianum* on bleeding patches. In severely infected palms, root feeding with hexaconazole (2 ml in 100 ml water) or carbendazim (5 g per 100 ml) and drenching of the basin with 40 L of 0.2 % hexaconazole or carbendazim.
2. Apply recommended dose of fertilizers N (500 g), P_2O_5 (320 g) and K_2O (1200 g) in two equal splits during June-July and December–January and provide irrigation (45 to 50 L per palm per day) during summer.
3. Apply neem cake (5 kg/palm) enriched with *Trichoderma harzianum* during September-October.
4. Avoid any type of injury to the trunks as wounds predispose the palms to infection. Care should be taken not to injure the stem base while ploughing the garden with tractor.

2.11 Harvesting

Usually 11- to 12-month-old nuts are harvested. Usually, the nuts are harvested at 6 to 10 times in a year.

References

Chowdappa P, Muralikrishna H and Rajesh MK 2017. *KALPA. CPCRI Newsletter,* **36** (4):1-20.

Dhanpal R and Thamban C 2007. *Coconut Cultivation Practices*. Extension Publication No.179, Published by Thomas GV, Director, CPCRI, Kasaragod - 671124, Kerala, pp.1-26.

Khan HH and Krishnakumar V 2002. Spices in coconut based cropping system. *In: Proceedings of the National Seminar on Strategies for increasing production and export of spices*, Calicut, 1-17.

Nimbolkar PK, Awachare C, Chander S, Husain F 2016. Multi storied cropping system in horticulture — A sustainable land use approach. *Int. J. Agric. Sci.*, **8** (55): 3016-3019.

Palaniswami C, Thomas GV, Dhanapal R, Subramanian P, Maheswarappa HP and Upadhyay AK 2007. *Integrated Nutrient Management in Coconut Based Cropping System*. Technical Bulletin No. 49, CPCRI, Kasaragod, Kerala, India, 24 p.

Roy S, Raj S, Choudhury M, Dey SK and Nazeer MA 2001. Intercropping of banana and pineapple in rubber plantations in Tripura. *Indian J. Natural Rubber Res.*, **14** (2):152-158.

Subramanian P, Thamban C, Vinayaka Hegde, Hebbar KB, Ravi Bhat, Krishnakumar V, Niral V and Josephrajkumar A 2018. *Coconut.* Technical Bulletin No.133, ICAR-CPCRI, Kasaragod, 56 pp.

Tandon HLS 1990. Integrated nutrient management for sustainable agriculture. *In: Proceedings of the International Symposium on Natural ReSources Management for a Sustainable Agriculture* (RP Singh, Ed.) Vol. I, 6th-10th February, 1990, New Delhi, pp 203-222.

Thomas GV, Subramanian P, Krishnakumar V, Gupta A and Chandramohanan R 2010. *Package of Practices for Organic Farming in Coconut.* Technical Bulletin No. 64. CPCRI, Kasaragod - 671124, Kerala, India. 28pp.

OUTCOME ASSESSMENTS

PART A

Answer the following questions true or false.

1. Coconut is essentially a subtropical plant. True/False
2. Stem bleeding is caused by bacteria. True/False
3. Irrigation is required by the coconut palm from November to May. True/False
4. Bud rot is caused by *Phytophthora palmivora*. True/False
5. The soil of the site should have a minimum depth of 2 m with good water holding capacity. True/False

PART B

Answer the following questions.

1. Stem bleeding disease is caused by fungus/bacterium.
2. Root (wilt) disease is transmitted by —.

3. Irrigation is required by the coconut palm from November to —.
4. Name some suitable intercrops for coconut orchard.
5. What are the major insect pests of coconut?

PART C

Write a brief note on each of the following.

1. Mother palm selection in coconut.
2. Coconut-based cropping system.
3. Land preparation and planting in coconut.
4. How can you manage rhinoceros beetle?
5. Nutrition management in coconut.

3

Production Technology of Arecanut

Botanical Name: *Areca catechu* L.
Family: Arecaceae / Palmae
Chromosome No: 32
Place of Origin: Philippines, Malay, Indonesia, Cochin, China and Malaya Peninsula
Common Name: Betel nut

3.1 Introduction

Arecanut *(Areca catechu* L.), commonly known as Supari or Betel palm or Arecanut palm, is cultivated primarily for its kernel obtained from the fruit which is chewed in its tender, ripe or processed form. It is an important cash crop of India. The economic produce is the fruit called 'betel nut' and used mainly for masticatory purposes. Its cultivation is restricted in South Western and North Eastern regions of the country. The chromosome number of *Areca catechu* L. was first determined as 2n=32 (Venkatasubban, 1945) and it belongs to family Arecaceae. It is one of the important commercial crops grown in parts of Karnataka, Kerala, Assam, Meghalaya, West Bengal, and Andaman and Nicobar Islands. The cultivation of arecanut has also been extended to other states like Tamil Nadu, Andhra Pradesh and Maharashtra. India is the largest producer and consumer of arecanut in the world holding 62% of the area and 60% of the production. The other major arecanut growing countries are Indonesia, Bangladesh, China, Myanmar and Thailand. Arecanut being a profitable plantation crop, it is important to understand the package of practices and seasonal operations to be followed in the nursery, in young gardens and in old plantations to get maximum returns. To get additional income in the adverse periods, concept of cropping system is very important. The exact origin of areca nut is not known till date but, arecanut is a palm which grows in much of the tropical Pacific, Asia, and parts of East Africa (Kumar, 2013).

Chemical composition of marketed arecanut depends on maturity of nuts since processed nuts are made from both green and ripe nuts. Major constituents are polyphenols, fat, polysaccharides, fibre and protein. Alkaloid, arecoline is present as significant constituent. Polyphenol content decreases with increasing maturity and hence tender nuts have better protection from infection compared to ripe nuts. Hardening of nuts during maturity is due to formation of polysaccharides. Arecanut has 1.5% of alkaloids such as arecoline, arecolidine, arecaidine, guvacine, isoguvacine and guvacolidine of which arecoline alone accounts for 0.24%. They possess anthelmintic property and are effective against tape worms and round worms. Tender nut contains tannins (about 38-47%), while ripe nuts have only 16-22%. Areca tannins obtained as by-product from tender nut processing can be utilized for dying cloths, leather, rope, for making black writing ink along with ferrous sulphate, as an adhesive in plywood manufacture and even as food colour (Chempakam, 1998).

3.2 Botany

Arecanut is an erect and unbranched palm. Height of palm depends on cultivar and environmental conditions. Stem has scars of fallen leaves in a regular annulated form. Arecanut palm has adventitious root system produced from basal bole. Primary roots give secondary and tertiary roots. Maximum concentration of roots is within a radius of 1m from bole and usually in top 60 cm of soil. Root hairs are absent. Stem becomes visible when palm is 3 years old. Mature stem is single, cylindrical, 30 m tall and 25-40 cm in diameter. It is green when young becoming greyish brown with age and ringed with leaf scars. Leaves are borne in terminal crown of about 2.5 m in diameter. One-year-old seedling has 4-5 leaves, which increase to 8-12 in adult palms. Six new leaves are produced per year. Flowering begins when the plants are 4-6 years old. Under best conditions, first inflorescence appears at 10th node at a height of 1.5-2 m from ground. Inflorescence is produced in axil of every leaf. It is a monoecious palm and its inflorescence is a spadix produced in the leaf axil and is completely enclosed in a sealed boat shaped spathe. The spadix is having a main rachis divided subsequently into secondary and tertiary rachis. Female flowers are confined to tertiary and distal end of the secondary rachis, while male flowers are produced on filiform branches arising below and beyond the female flowers. Both female and male flowers are sessile, with two whorls of perianth. Male flowers range from 15,000 to 50,000 and are arranged in pairs of two rows along upper part of thin branches. Male phase in arecanut lasts for 25-46 days. Spadix of a grown-up palm bears up to 600 female flowers borne on thickened bases of secondary or tertiary branches. Female flower buds start opening after all male flowers are shed. Anthesis is from 2 A.M. to 10 A.M. Female phase extends for 3-10 days. Maximum receptivity is between 2nd and 4th day of opening. Middle aged palms have higher stigmatic receptivity than young and old palms. The fruit is a monolocular, one seeded berry and it consists of a fibrous outer husk, enclosing a single seed. It is a cross-pollinated crop and fruit set normally varies from 12.0 to 40.0 percent and the time taken from full bloom to maturity of the fruit ranges from 35 to 47 weeks.

3.3 Climate and Soil

The arecanut can be grown in a variety of climatic and soil conditions. It grows well from almost sea level up to an altitude of 1000 metres in areas receiving ample and well circulated rainfall or under irrigated conditions. Arecanut thrives well in humid areas protected well against sunburn and heavy wind. Aligning the rows in north-south direction with a deviation of 35° towards south-west lowers the incidence of sun scorch. Since the areca palm does not withstand either water logging or drought, the site selected should have adequate facility of water for irrigation and the soil should have proper drainage facilities. It can grow from Western Plains and Ghats to North Eastern Hills. Arecanut grows in a wide range of temperature between a minimum of 4°C and a maximum of 40°C. However, the palm flourishes well within a temperature range of 14°C to 36°C. It grows well in areas where annual rainfall may go up to or even more than 4500 mm. But it also survives in low rainfall areas having 750 mm annual precipitation. During prolonged dry spell palms should be irrigated. Though, arecanut is grown under different agro-climatic conditions, it is very sensitive to extreme climatic conditions (Bhat and Abdul Khader, 1982). High humid conditions provide congenial conditions for the rapid spread of diseases like fruit rot, bud rot etc.

It can grow in laterite, red loam and alluvial soils. The soil should be deep and well drained. In plain region, it is cultivated in fertile clay loam soils. Arecanut is mostly grown in red lateritic soils in Western Ghats region, West Bengal and Assam, and clay loams in plains of Karnataka (Mohapatra and Bhat, 1982). It can sustain soil pH range from 5.2 to 7.0. To reclaim acid soil, lime is applied @ 0.5 kg/palm/year once in two years and incorporated by forking during April-May (Balasimha and Rajagopal, 2004).

3.4 Choice of Cultivars

Systematic evaluation of exotic and indigenous accessions of Arecanut through selection and hybridization at CPCRI, Regional Station, Vittal, Karnataka resulted in release of cultivars with high yield and desirable characteristics.

Table 3.1: Cultivars with desirable characteristics

Sl.No.	Characteristics	Cultivars
1	High yielding	CPCRI Varieties, Calicut-17, Sirsi Arecanut Selection (SAS-1)
2	Early bearing	Mangala
3	Higher number of fruits	Tirthahali, Devadarshini
4	Better quality nuts	Shreevardhan, Sweet areaca
5	Large fruit size	South Kanara Local, Kahikuchi, Calicut-17, Sreemangala
6	Regular bearing	South Kanara Local
7	Semi-tall	Mangala, Shreevardhan, Swarnamangala
8	Dwarfness	Hirehali Dwarf, VTLAH-1. VTLAH-2
9	Uniformity	Mohitnagar
10	Tender nut	Shreevardhan, Tirthahali, Hirehali Tall

Source: CPCRI Newsletter

Table 3.2: Arecanut growing areas with suitable cultivars

Sl.No.	Growing areas	Cultivars
1	All Arecanut growing areas	CPCRI Varieties
2	North Kerala, South Karnataka	South Kanara Local, Kasaragod Local
3	Malnad areas of Karnataka	Tirthahali
4	Shimoga and Uttar Kannada	Sagar
5	Coastal Maharashtra and Karnataka	Shreevardhan
6	Tamil Nadu	Mettupalyam Local, Mohitnagar
7	Assam, West Bengal	Kahikuchi Tall, Mohitnagar
8	Andaman and Nicobar Islands	Calicut-17

Source: CPCRI Newsletter

Genetic resource programme in arecanut was undertaken at the CPCRI Regional Station, Vittal, Karnataka. Germplasm holding consists of five species, *viz*., *A. catechu, A. triandra, A. macrocalyx, A. normanbyii* and *A. concinna,* and two genera, *Acctinorhytis* and *Pinanga dicksonii*, and 133 accessions. Screening of germplasm accessions led to release of following high yielding cultivars (Peter, 2002):

MANGALA: It is a selection from VTL 3, an introduction from China in the year 1957. It is semi-tall, early bearing with quicker stabilisation of yield and producing nuts with good chewing and market quality. Flower initiation occurs at 36-40 months after planting Nuts are small and round in shape. It produces 10 kg ripe nuts / palm / year which gives 2 kg *chali*. It produces mean yield of 2.90 kg dry kernel /palm/year and 3700 -3900 kg dry kernel/ha/year. Its potential yield is 4.50 kg dry kernel/palm/year and 6000-6300 kg dry kernel/ha/year. (Annual Report. 1971- 72. Central Plantation Crops Research Institute, Kasaragod, India,166 pp).

SUMANGALA: It is an introduction from Indonesia. Palm is tall with partially drooping habit. Flower initiation occurs at 42-46 months after planting. Nuts are medium sized, deep yellow to orange in colour and oblong to round in shape. It gives an average yield of 17.25 kg ripe nuts / palm / year at 10th year. It produces mean yield of 3.28 kg dry kernel/palm/year and 3900 -4350 kg dry kernel/ha/year with a potential yield of 5.60 kg dry kernel/palm/year and 7400-7700 kg dry kernel/ha/year. It is relatively tolerant to water stress conditions. No major diseases and pest attacks observed under field conditions. (Annual Report 1984-85 of Central Plantation Crops Research Institute, Kasaragod, Kerala, India, 213 pp).

SREEMANGALA: This cultivar was selected from an introduction from Singapore. Habit, flowering and fruit characters are similar to Sumangala. Flower initiation occurs at 44 to 48 months after planting. Nuts are round and bold. It gives an average yield of 15.63 kg ripe nuts/ palm / year. It produces mean yield of 3.18 kg dry kernel/palm/year and 4240-4500 kg dry kernel/ha/year with a potential yield of 5.40 kg dry kernel/palm/year and 7100 -7300 kg dry kernel/ha/year. (Annual Report 1984-85 of Central Plantation Crops Research Institute, Kasaragod, Kerala, India, 213 pp).

SWARNAMANGALA: Evaluation of exotic accessions and selection for economic traits resulted in identifying a high yielding variety VTL–12 (Saigon). It is a tall cultivar with homogeneous population. Flower initiation occurs at 46 months after planting. Bunches are well spaced; nuts are bold and heavier with high recovery of chali/dry kernel (26.40%). Average yield of this palm is 3.88 kg chali/palm/year (Anon, 2007). It produces mean yield of 3.88 kg dry kernel/palm/year and 5320 kg dry kernel/ha/year with a potential yield of 5.50 kg dry kernel/palm/year and 7700 kg dry kernel/ha/year. No major disease/pest attacks are observed under field conditions. Prophylactic pests' measures are required for fruit rot disease, in endemic tracts. (Annual Report 2005-06. Central Plantation Crops Research Institute, Kasaragod, Kerala, India, 128 pp)

MOHITNAGAR: This is an indigenous cultivar from West Bengal. Important feature of this cultivar is its superior uniformity. Flower initiation occurs at 48 months after planting. Bunches are well-spaced and nuts are loosely arranged on spikes which help in uniform development enabling efficient plant protection measures. Nuts are round and medium in size. Early stabilization of yield and high annual yield potential of 3.7 kg *chali*/palm (15.08 kg ripe nuts) was also noticed. It produces mean yield of 3.67 kg dry kernel/palm/year and 5030 kg dry kernel/ha/year with a potential yield of 5.50 kg dry kernel/palm/year and 7540 kg dry kernel/ha/year. It is relatively tolerant to water limited conditions. (Annual Report 1990 -91. Central Plantation Crops Research Institute, Kasaragod, Kerala, India, 142 pp.)

KAHIKUCHI: It is mainly cultivated in Assam and Meghalaya. Palm is tall with medium thick stem with consistency in yield. It produces a mean yield of 3.70 kg dry kernel/palm/year and 5073 kg dry kernel/ha/year with a potential yield of 5.30 kg dry kernel/palm/year and 7000-7200 kg dry kernel/ha/year. Economic yield can be realized up to 45 years depending upon the management. Fruits are orange colour, bold having round shape nuts. Flower initiation occurs at 50-52 months after planting. It is recommended for cultivation only under irrigated conditions. (Annual Report 2008-09. Central Plantation Crops Research Institute, Kasaragod, Kerala, India, 145 pp).

SAMRUDHI: This cultivar is recommended for Andaman and Nicobar Islands. It is tall with longer internodes and crown. The prominent feature of this cultivar is its consistent and high yield potential (18.89 kg ripe nuts/palm/year with a chali yield of 4.34 kg/palm) having well placed bunches with round and bold nuts.

SAS-1 (SIRSI ARECANUT SELECTION-1): It is released from University of Agricultural Science, Dharwad. It is recommended for hill zone of Karnataka. The cultivar is tall with compact canopy and is a regular bearer. Nuts are round and even sized and closely arranged on compact bunches. It is suitable for both tender and ripe nut processing. It has high curing percentage and yields 4.60 kg chali/palm/year.

MADHURAMANGALA: It is mainly grown for dry kernel and tender nut processing. It is recommended for cultivation in Karnataka and Maharashtra. Flower initiation occurs at 42 months after planting. Economic yield can be obtained up to 35 years, depending upon the management. Fruits are orange to yellow colour, oval and round shape, with medium sized nuts. There is high chali recovery (26 %) from fresh nuts. It produces a mean yield of 3.54 kg dry kernel/palm/year and 4500-5000 kg dry kernel/ha/year, 2.95 kg dry tender processed nuts/palm/year and 3800-4500 kg dry tender processed nuts/ha/year. It is recommended for cultivation only under irrigated conditions. (Annual Report 2011-12. Central Plantation Crops Research Institute, Kasaragod, Kerala, India, 128 pp.)

SHATAMANGALA: It is recommended for dry kernel and tender nut processing and is mainly grown in areas of Karnataka and Gujarat. Flower initiation occurs at 40 months after planting. It produces a mean yield of 3.96 kg dry kernel/palm/year and 5000 kg dry kernel/ha/year; 3.26 kg dry tender processed kernel/palm/year and 4500 kg dry tender processed kernel/ha/year. Palms are semi-tall having medium thick stem, shorter internodes, with partially drooping crown. Economic yield can be obtained up to 45 years depending upon the management. Fruits are orange in colour, round shaped having medium size nuts with high recovery (26.8%) of chali from fresh nuts. It is recommended for cultivation under rainfed and irrigated conditions. (KALPA CPCRI Newsletter 35 (2) April-June 2016, Central Plantation Crops Research Institute, Kasaragod, Kerala, India, pp. 2.).

SOUTH KANARA: It is largely grown in South Kanara district of Karnataka and Kasaragod of Kerala. The palms are regular bearing in habit with large sized nuts. It produces a yield of about 7 kg ripe nuts/palm/year giving 1.5 kg chali.

SREEVARDHANA: It is grown in coastal Maharashtra. Nuts are oval shaped with marble white kernel and tastier endosperm which are rated as the best quality. Yield is comparable to 'South Kanara'.

VTLAH 1 (Hirehalli Dwarf × Sumangala): This hybrid is dwarf in nature with well-built stem with superimposed nodes, reduced canopy size, well spread leaves, partial drooping crown, medium sized oval to round and yellowish orange-coloured nuts. Early stabilization in yield and high recovery of chali (26.45%) are the remarkable features of this hybrid. Average chali yield of this hybrid is 2.54 kg/palm/year. The dwarf hybrid with high yield potential will directly benefit the growers by way of enhanced returns and reduced cost of various cultural operations like harvesting, spraying and also low damage to palms due to sun scorching and heavy wind (Anon, 2007).

VTLAH (Hirehalli Dwarf × Mangala)

It is an early bearer, dwarf with partially drooping crown, well spread leaves, sturdy stem, medium size, oval shape and yellow to orange-coloured nuts. It reduces cost of cultivation, harvesting and spraying easy. Average yield (Chali/ palm): 2.54 Kg. It is released for coastal Karnataka and Kerala.

3.5 Propagation

It is mainly propagated through seed. Being a perennial seed propagated crop, much care is to be taken on production and selection of planting materials. Production of good quality planting materials in arecanut at different stages includes selection of mother palm, selection of seed nuts, adoption of proper nursery techniques and selection of seedlings.

3.5.1 Selection of Mother Palms

Select mother palms showing earliness in bearing and high percentage of fruit set. It is preferable to select palms with shorter internodes, more number of leaves on crown and producing at least 4 bunches in a year (Peter, 2002). According to Bavappa and Ramachander (1968) age at first bearing and regular bearing habit are two important characters to be considered for selection of mother palm. Other characters to be considered for selection of mother palms are larger number of leaves on the crown, shorter internodes and high fruit set. In the erstwhile Mysore state, the seed nuts were collected from the palms aged between 25 to 30 years (Aiyer 1966).

3.5.2 Selection of Seed Nuts

Select fully tree ripe nuts from middle portion of middle bunch on tree. Heavier nuts gave higher percentage of germination and produced vigorous seedlings. Selected bunches are lowered by means of rope and sown without delay. Nuts selected from the middle portion of the middle bunch neither produced better seedlings nor trees with better performance (Anonymous 1963). The heavier nuts gave better germination (96% against 87% from lighter nuts), produced seedlings with greater vigour and more number of quality seedlings (Bavappa and Abraham, 1961).

3.5.3 Sowing

The selected seed nuts should be sown soon after harvest in nursery beds prepared under shaded areas with stalk end up and with a closer spacing of 5-6 cm. The seed nuts have to be covered with sand and irrigate daily. Germination starts in 45 days and continues up to 3 months. Good quality seeds give 90-98% germination. The seedlings are retained in primary nursery for about six months. The secondary nursery beds of 150 cm width and of convenient length are prepared for transplanting the sprouts. The sprouts are transplanted at a spacing of 30 cm × 30 cm with the onset of monsoon. The spacing with which the sprouts are planted has significant influence on the growth of seedlings (Bavappa and Mathew, 1960). It was found that sprouts planted at 45 cm are more vigorous than those planted at 15 cm. Partial shade to the seedlings can also be provided during summer by pandal or growing banana. Care should be taken to drain the nursery beds during the monsoon and to irrigate them during the dry months. Weeding and mulching should be done periodically. Seed nuts can also be sown in polythene bags (25 × 15 cm size and 150-gauge thickness) after filling the bags with potting mixture containing 7 parts of loam or top soil, 3 parts of dried farm yard manure and 2 parts of sand. The seedlings will be ready for transplanting to the main field when they are 12 to 18 months old. Seedlings having 5 or more number of leaves are selected.

3.5.4 Selection of seedlings

Good seedlings of 12-18 months old are selected for planting in the main field. Seedlings with maximum number of leaves and minimum height are to be selected for planting. The seedlings selected should have maximum number of leaves (five or more) and minimum of 90 cm height (Bavappa, 1970). The seedlings have to be uprooted with a ball of earth adhering to roots. Covering the base with ball of earth with plastic bag will help in keeping the seedlings in good condition for long distance transport (Anonymous, 1964).

3.6 Planting

Arecanut palm is very delicate and cannot withstand extremes of temperature and exposure to direct sun. Tall, quick growing shade trees are planted on southern and western sides of plantation to provide protection from sun scorching. Dig pits of size 75-80 cm^3 at a spacing of 2.7 m × 2.7 m with North-South alignment. A spacing of 2.7 m × 2.7 m is recommended for arecanut planting. When arecanut is planted as a mixed crop with other crops, wider spacing of 3.3 m × 3.3 m will be optimum. Dwarf areca cultivars and hybrids may be grown in 2.2 m × 2.2 m spacing. Plant seedlings in centre of pits and cover with soil up to collar level and press around. Planting is done during May-June in well drained soils and during August-September in clayey soils provided water logging condition should be avoided. Banana can be planted between rows to provide shade in initial stages up to 4 to 5 years or else artificial shade can be provided with coconut leaves or areca leaves during hot weather period / starting from October. Stems of palms can be protected from sun scorch by covering with dry leaves and spathe of arecanut or white opaque polythene film. In areas where South-West monsoon is severe, planting in the month of September-October is recommended. In other areas planting can be done in the months of May-June. Mulching should be done immediately after planting, before end of monsoon, to avoid drying up of top soil.

3.7 Nutrition

Sufficient provisions of plant nutrients in the soil right through the life of the crop are crucial to get high yield. Hence, an annual application of 100: 40: 140 g of NPK per tree in the form of fertilizers and 12 kg each of FYM or compost or cattle manure per bearing palm is recommended. Apply 1/3rd dose during first year, 2/3rd dose during second year and full dose from third year onwards. Under irrigated conditions, fertilizers are applied in two equal split doses during September-October and February. Under rainfed conditions, second dose is applied during March-April after receipt of summer rains. Manures and fertilizers are applied in basins around palm dug to a depth of 15-20 cm and 0.75-1.00 m radius from base of palm and covered with soil. Second dose is applied after weeding and forked in. The results of a study in which organic manure (green leaf, cattle manure, bone meal and wood ash) and/or inorganic

NPK were applied at equivalent major nutrient rates, showed that the nutrient source did not influence plant growth and crop yield. The nutrient content of the soil was increased after the application of different treatments indicating any form of fertilizer can be used for arecanut (Bhat and Mohapatra, 1989).

3.8 Water Management

Arecanut cannot withstand drought for a long time. Being a perennial crop, once affected by water stress, it may require two to three years to regain its normal vigour and yield. Arecanut, being the most profitable cash crop, irrigation has positive and significant effect on economics of the crop (Dinesh Kumar and Mukundan, 1996). When there is shortage of water, drip irrigation can be practised. At Vittal, irrigation intervals of 5 and 10 days were found superior over 15 and 20 days (Abdul Khader *et al.*, 1985). When irrigation was scheduled based on IW/CPE ratio, it was found that irrigation of 30 mm of water when the CPE is 30 mm (IW/CPE ratio of 1) is optimum (Yadukumar *et al.*, 1985). In Kerala, arecanut plantations are irrigated during dry months once in seven or eight days during November-December and once in six days during March, April and May. Saved water by adopting drip irrigation can be effectively used for irrigation of other crops/ intercrops and depletion of ground water can be checked. For pre-bearing arecanut palms 50 per cent of recommended nutrient is sufficient when it is given through drip irrigation thus saving considerable amount of fertilizer cost (Sujatha *et al.*, 2002). Adequate drainage should be provided during monsoon since the palms are unable to withstand water logging. Drainage channels should be 25 to 30 cm deeper than the bottom of the pits to drain excess water from the plot. Mulching is very important in arecanut because of highly porous nature of soil and greater seepage losses. Various plastic and organic mulching materials were effective.

3.9 Intercultural Operations

The main principle of intercultural operation is to loosen the soil and to rebuild the soil fertility after the heavy rains during monsoon. In general, the cultivation has found to increase the yield by 10-20%. In light soils digging the soil can be done once in two years. But in heavy soils digging has to be done every year. Clean cultivation has found to give better yield with lesser weeds.

3.10 Cropping System

3.10.1 Intercropping

It is a system where annual or biennial crops are grown with arecanut. Around 60% of light is intercepted by an adult palm. By growing intercrops, level of light interception can be increased to 95%. Only 35% space is used by arecanut palm and rest 65% area is available for intercropping. Hence, crops like paddy, sorghum, cowpea, vegetables and yams can be grown in the interspaces. For these crops, agro-techniques should be followed as per sole crop recommendation. Pits or trenches are taken for dioscorea, elephant foot yam, banana and pineapple. Crops like ginger, turmeric, arrowroot, chillies etc. are planted in raised beds of convenient size. Crop selection should be done according to local preference and market. Banana is a popular intercrop and gives income during early years of planting of arecanut and throughout the growth period as well. Of late, medicinal and aromatic plants can be grown as a source of income. With procurement facilities like buy back schemes, the potential returns are quite high.

Multiple cropping in arecanut as a productive land use system through the use of interspaces has been practised (Sannamarappa and Muralidharan, 1982). The long pre-bearing age of arecanut has prompted farmers to grow different annual or biennial crops for economic sustainability. This initial period of 5-6 years is ideal for growing short duration crops. In later years, as the arecanut canopy increases in height, mixed cropping with other shade tolerant perennial crop species is used. Thus, there is an excellent opportunity for a temporal and spatial distribution of crop species in arecanut gardens. A number of annual crops like paddy, sorghum, cowpea, vegetables and yams are grown as intercrops of arecanut palms (Shama Bhat, 1974; Thomas,1978).

3.10.2 Mixed Cropping Systems

Growing of perennial crops as associated crops in arecanut garden is called mixed cropping system. Several combinations of crops can be grown along with arecanut. Black pepper, banana, cocoa, cardamoms are excellent crop for mixed cropping with arecanut. Mixed cropping in arecanut plantation has promoted more growth and yield of main crop of arecanut as indicated by increased number of leaves (fronds) and increased yield per palm compared to sole crop (Girish *et al.* 2003).

Table 3.3: Efficient cropping models for different arecanut growing regions

Sl No.	Growing regions	Cropping models
1	Maidan parts of Karnataka	Arecanut + Black Pepper + Cocoa
2	Kerala and coastal parts of Karnataka	Arecanut + Black Pepper + Cocoa + Banana
3	North Bengal	Arecanut + Black Pepper + Banana or Arecanut + Black Pepper + Acid lime
4	Wynad of Kerala and Uttara Kanada of Karnataka	Arecanut + Cardamom Arecanut + Coffee

3.10.3 High Density Multiple Cropping System

It is a system where more than two crops are grown simultaneously with arecanut. The choice of component crops mainly depends upon its ability to grow under the shade of arecanut palm, to withstand high rainfall. Arecanut + Cocoa + Banana + Black pepper is the widely adopted system. High density multispecies cropping system in arecanut typically comprises of black pepper, cocoa/clove, pineapple, coffee and banana occupying different vertical air spaces. The success of involving multispecies cropping system depends on the relative shade tolerance of component crops. More shade tolerant species are desired at still lower vertical heights. Among the six crops studied, cocoa, clove, coffee, banana and pepper are suitable for multi-storeyed cropping (Bavappa, *et al.*, 1986).

Sl No.	Crop	Pit size (cm)	Spacing (m)	Fertilizer (N: P: K) (g/plant)
1	Banana	50 × 50 × 50	2.7 × 5.4	160:160:320
2	Black Pepper	50 × 50 × 50	2.7 × 2.7	100:40:140
3	Cocoa	50 × 50 × 50	2.7 × 5.4	100:40:140
4	Acid lime	50 × 50 × 50	2.7 × 5.4	300:250:500
5	Betle vine	50 × 50 × 50	2.7 × 2.7	100:40:140

3.11 Green Manuring and Cover Cropping

Screening of suitable green manure crops for arecanut garden showed that *Pueraria javanica* and *Mimosa invisa* are good green manure-cum-cover crops in arecanut gardens because of their good manure yield and nutrient addition capacity. *Mimosa invisa* and *Calopogonium. mucunoides* gave the best improvement in soil organic carbon status (Mohapatra *et al*, 1970). *Sesbania* was a good crop which can withstand water logging and drought. *Calopogonium* and *Pueraria* may be used as cover crops in hilly slopes also to prevent soil erosion. The seed requirement per hectare for *Mimosa, Stylosanthes, Calopogonium* and *Pueraria* is 15 kg, 9 kg, 11 kg and 11 kg respectively. The cover crops may be sown in the month of April-May and the green matter may be cut and applied to arecanut palms at the time of second dose of fertilizer application during September-October.

3.12 Crop Protection

The productivity of palm is affected by a number of pests, diseases and disorders. Some of the diseases like fruit rot, yellow leaf disease, basal stem rot and inflorescence dieback are economically important since they cause damage to the product as well as the crop. The other diseases of lesser importance are bud and crown rots, bacterial leaf stripe, stem bleeding, seedling diseases and the disorders like crown choke, nut splitting, sun scorch and stem breaking etc. The yellow leaf disease caused by Phytoplasma is a slow decline disease, which affect the productivity of the palm as well as the quality of nuts.

3.12.1 Insect-Pests

Arecanut is attacked by different insect pests. The pest infests all parts of the palm, *viz.*, stem, leaves, inflorescence, roots and nuts. Among these, mites, spindle bug, inflorescence caterpillar, root grubs and pentatomid bug cause considerable economic loss to the crop.

1. Mites *(Raoiella indica, Oligonychus indicus)*

The mites are widely distributed in all areca growing regions of India. They are polyphagous and found to occur on other palms also. Adults and young ones suck the lower surfaces of the leaves, causing them to turn yellow and bronzed appearance. In the case of infestation all the leaves in the seedlings are affected causing yellowing and often death of the seedling. In older palms infestation starts in the lower whorl of leaves and as the population increases, it spreads to the inner whorl. With onset of hot weather from April-May, they become more active and dangerous (Patel and Rao, 1958). The plantation under drought stress and nurseries are more prone to mite infestation. The pest incidence is at a lower level under well-irrigated and partially shaded conditions. The mite population declined with the onset of monsoon.

It could be controlled by spraying dicofol (0.05 %) or Kelthane (0.1%) (Anonymous 1967, Kantha *et al.*1963).

2. Spindle Bug *(Carvalhoia arecae)*

Both the nymphs and adults suck sap from the tender spindle and leaves. The bug penetrates the stylet into the tissues by bending the rostrum and starts feeding. The feeding injury is caused on the lamina and petiole. The affected leaves show dry brown patches. The spindle infested by the bug shows typical linear brown lesions. Severely infested spindles fail to open fully.

Filling the innermost two leaf axils around the spindle with granular systemic insecticides like Phorate 10 G (or Thimet 10 G) at an interval of three months was suggested (Abraham *et al.*, 1976). Spraying Fish Oil Rosin Soap (FORS) (1 kg in 80 litres of water), quinalphos (1ml in one litre of water) could control the pest (Anonymous, 1964).

3. Inflorescence Caterpillar *(Tirathaba mundella)*

Caterpillars feed on the flowers and compress the inflorescence into a wet mass of frass with silky threads. Adult moths lay eggs in holes made in spathe by slugs and earwigs. On hatching, caterpillars expand entry into inflorescence through these holes and feed on flowers inside and transform it to a wet slimy mass with silken threads. Affected spadices do not open.

If the damage is only partial, the affected portions are removed and the inflorescence is sprayed with 0.125% malathion (Anonymous, 1971). Damage can be reduced indirectly by control of slugs by mechanical methods as well as by poison baiting with metaldehyde.

4. Root Grub *(Leucopholis lepidophora)*

Root grubs attack and feed on roots of both young and old palms. Due to root feeding, leaves turn pale yellow and yield gets reduced. The plantation should be clean, well-drained and free of weeds which reduces its attack.

Soil should be loosened around base of palm to a depth of 10-15 cm and drenched with chlorpyriphos (0.04%) suspension twice, in May and September. The application should be repeated for 2 or 3 years for a complete eradication of pest. Soil application of Phorate 10 G around palms is also effective.

5. Pentatomid Bug *(Halyomorpha marmoreal)*

Adult and young bugs suck sap from endosperm of tender nuts and cause premature shedding. Dropped nuts have one or more marks on surface. Adults are bronze coloured with brown spots and in young stages, they are black with white spots on leg. When tender nuts are not available, insect migrates to other hosts like cowpea and bitter gourd and so these crops should be closely watched and bug when noticed should be mechanically removed and destroyed. Spraying with chlorpyriphos @ 2ml/plant on bunches is found effective.

3.12.2 Diseases

1. **Koleroga (Mahali or Fruit Rot)** (*Phytophthora arecae*)

The symptoms are appearance of water-soaked lesions on the nut surface near the perianth and later it spreads over the other parts. Infected nuts shed without perianth. The disease makes its first appearance after first monsoon rain and extends throughout monsoon period. First symptom is appearance of water-soaked lesions on nut surface near the calyx. Later the patches enlarge and nuts become dark and are shed in large numbers. Fallen nuts soon develop whitish mycelia mass all over. Nuts of all ages are attacked and if not checked, attack the crown causing leaves and bunches to wither. Sometimes, infected nut may not be shed and remain mummified in bunches. Such type of infection is known as 'dry mahali'.

Spray Bordeaux mixture 1% on all bunches three times in a year, one just before the onset of South-West monsoon and the rest at 40 days intervals. If monsoon season is prolonged, give a third spray. All fallen and infected nuts should be removed and burnt.

2. Anabe or Foot rot *(Ganoderma lucidum)*

This disease is caused by a fungus *Ganoderma lucidum*. The disease is severe in neglected, ill- drained and over-crowded gardens (Venkatarayan, 1936). It is more common in neglected gardens. Symptoms of disease are similar to that of drought. Initial visible symptom is yellowing of outer whorl of leaves, which gradually extends to inner whorls followed by wilting and drooping. Development of inflorescence and nuts are arrested and nuts already formed are shed. At later stages, weakened crown topples off leaving a bare trunk. All around base of palm, brownish patches appear which exude a brown liquid. Interior of stem at basal region is discoloured and rotten, emitting a foul smell. Infection extends to roots and they get discoloured, brittle and dry. Fungal invasion interrupts uptake of water and nutrients by palm, leading to yellowing and wilting.

Isolate affected palms by digging trenches of 60 cm deep and 30 cm wide around, one metre away from base and drench with Captan (0.3%). Palms in the early stage of disease may be root fed with 125 ml of 1.5% Calixin solution at quarterly intervals i.e., during March, July, October and January. Soil drenching with Bordeaux mixture (1%) before planting healthy seedling should be followed. Growing of collateral hosts of fungus like *Delonix regia* and *Pongamia glabra* should be discouraged.

3. Yellow Leaf Disease (YLD)

Yellow leaf disease is the most serious disease affecting arecanut, though it is restricted to Kerala and Karnataka. Phytoplasma is found associated with the disease. The plant hopper *Proutista moesta* acts as vector in spread of disease.

Disease is characterized by yellowing of leaves followed by necrosis. Withering of tips starts and gradually spreads to older portions of leaf. Freshly formed leaves grow shorter and their laminas show unequal growth and flaccidity. In a few cases, wilting and shedding of leaves are also observed. Nuts are reduced in size, shrivelled with their kernels often turning black and there is severe reduction in yield. Stem of affected palms becomes spongy and friable, conducting strands getting destroyed. In advanced stages, stem breaks off at top. Rotting of roots is also observed. Chlorophyll content is reduced. Water logging is a predisposing factor for incidence of disease. Lack of balanced nutrition and unscientific cultivation practices make palm susceptible to disease.

Diseases caused by Phytoplasma are not curable by application of conventional plant protection chemicals. It is useful to adopt proper management practices to get additional income from affected gardens (Reddy *et al.*, 1987). Soil application of NPK and lime with or without zinc sulphate significantly reduced foliar yellowing (Dastagir, 1965). Since deficiency of phosphorous was observed in diseased gardens, application of additional dose of one kg of super phosphate to affected palms and palms given additional dose of P delayed symptom expression on areca seedlings (Anonymous, 1983). Though disease cannot be cured, general health of affected palm and economic yield could be improved in a diseased garden and following measures are suggested:

1. Adequate drainage should be provided especially during monsoon.
2. Phytosanitary and plant protection measures should be adopted to control *anabe*, bud rot, spindle bug and mite infestation.
3. Sun scorching of stem should be avoided by covering with arecanut leaves or painting with lime slurry.
4. In addition to application of NPK fertilizers as per schedule apply rock phosphate 160g per palm and lime and zinc @ 85g per palm.

4. Bud Rot *(Phytophthora palmivora)*

Affected spindle appears yellow, later changing to brown and finally the whole spindle rots. Remove and destroy affected spindle and leaves. In early stages of infection, scoop out affected rotten tissues by making longitudinal side splits and apply Bordeaux paste on the exposed healthy tissues or drench crown with 1% Bordeaux mixture.

5. Stem Bleeding

Palms in the age group of 10-15 years are more susceptible to this disease. Symptoms appear on the basal portion of the stem as small discoloured depression. Later, these spots combine together and cracks develop on the stem leading to disintegration of the fibrous tissues inside. With the progress of the disease a brown exudate oozes out of these cracks. High water table predisposes the palm to this disease.

Improvement of drainage and root feeding of 125 ml Tridemorph (1.5%) is suggested as control measure against this disease. Scooping out affected portions and application of coal tar or Bordeaux paste is effective in reducing incidence.

3.13 Physiological Disorders

3.13.1 Band or Hidimundige

Band is a physiological disorder of palm due to adverse environment of the particular spot where palm is standing. First visible symptom is reduction in leaf size which turns brittle and crinkled with wavy margins. As disease advances, there is reduction in internodal length, formation of small bunches and tapering of stem. Crown exhibits a rosette shape due to failure of natural opening of leaves. Bunches formed will be small and malformed. Roots are poorly developed, crinkled and brittle. Poor drainage, low soil fertility or environmental factors are reported as possible causes of malady. Disease is reported from Maharashtra and Karnataka.

Good soil management, improvement of drainage and incorporation of copper sulphate and lime to soil could check disease effectively.

3.13.2 Nut Splitting

Nut splitting is more a physiological disorder than a pathological problem and is commonly occurred in areca growing areas. It is seen in well grown young and healthy palms. Growth of pericarp does not keep pace with development of kernel inside and thus causes splitting up of pericarp. Split nuts drop and infection of exposed kernel renders them useless.

Splitting is due to excess flow of cell sap into inflorescence in very healthy palms. Hence, checking of excess flow either by making some deep wound at base of spadix or jamming cells at base when nuts are half grown prevents splitting. Sudden application of water after a period of drought also results in nut splitting. Potassium deficiency is also reported as a probable cause of this malady. Application of potash fertilizers and spraying of borax (2g / litre of water) during early stages of disease reduce splitting.

3.13.3 Sun Scorch or Stem Breaking

This disorder resulted from prolonged exposure of palms to severe solar radiation. Palms exposed to South West sun are more prone to stem breaking.

Symptoms appear as golden yellow splits on exposed side of stem, which turn dark brown and subsequently form longitudinal fissures. Further colonisation by saprophytic fungi weakens stem and finally breaks. Growing of fast-growing trees in the South-West side of garden, protecting stem with dry areca leaves, trailing pepper vines on stem, and adopting a suitable alignment for planting are suggested to minimize disease.

3.14 Harvesting and Yield

The stage of harvesting varies with the regions depending on the consumer market. Two types of final product are seen in arecanut. One is prepared out of immature nuts and the other from ripe nuts. Arecanut palms start flowering from 3-4th year of planting. December-March is the main flowering season and harvesting period is from June-July for tender nuts and November-March for ripe nuts. Nuts are harvested at 45-50 days interval in three pickings. Nuts are processed both at tender as well as ripe stage. Harvesting is done either by climbing palm or by using a long bamboo with a sharp sickle or hook attached to end.

More than 10 kg of ripe nuts per palm at the 10th year is considered as normal yield in any plantation.

3.15 Postharvest Handling

3.15.1 Chali or Kottapak (White nuts)

The most popularly traded form of arecanut is the dried whole nut known as Chali or 'Kottapak'. Fully ripe, nine months old fruits having yellow to orange red colour is the best suited for the above purpose. Ripe fruits are dried in the sun for 35 to 40 days on dry levelled ground. For drying and dehusking, sometimes fruits are cut longitudinally into halves and sun dried for about 10 days, then the kernels are scooped out and given a final drying. Optimum moisture content is around 12%. Inadequate drying results in fungal infection and in a poor-quality product. Mechanical driers are also used to make chali. Drying takes 60-70 hours over a period of 7-8 days at 45-75°C. Dehusking can be done using a manually operated arecanut dehusker developed by CPCRI, Kasaragod. About 40 kg chali can be made within a period of 8 hour.

3.15.2 Kalipak (Red nuts)

Kalipak is another important processed product of arecanut. Kerala and Karnataka are main producers of Kalipak. Tender nuts of 6-7 months are dehusked, cut into pieces, boiled with water or a diluted extract from previous boiling, coated with kali and dried. Kali coating can be repeated 3-4 times to get a glossy appearance. Kali is the concentrated extract obtained after 3-4 batches of boiling. During preparation of kalipak, tannin content gets reduced substantially. A well dried product with a dark brown colour, glossy appearance, crisp chewing feel, well-toned astringency and absence of over mature nuts are rated superior.

3.15.3 Scented Supari

Scented supari is made both from chali and Kalipak. Chali supari is more popular. Dried nuts are broken into bits, blended with spices and flavour mixture and packed in butter paper or aluminium pouches. Instead of raw species, essential oils are used for easy blending, with coconut gratings to avoid microbial growth. Saccharin is occasionally used for sweetening and rose essence and menthol are commonly used for flavouring. A product blended with cashew nut bits called kaju-supari has been developed. Scented suparis popular in north and central India are of two types: the one made from chali and the other from kalipak. The former is more popular.

References

Abdul Khader KB, Yadukumar N and Bhat KS 1985. Irrigation requirement of arecanut (*Areca catechu* L.). *Proc. Silver Jubilee Symposium on Arecanut Research and Development.* CPCRI, RS, Vittal. Dec. 13-14, 1982. pp 27-32.

Abraham VA, Sathiamma B, Abraham KJ and Kurian C 1976. Control of arecanut spindle bug (*Carvalhoia arecae*) using granular insecticides. *J. Plantn. Crops*, **4**: 24-25.

Aiyer AKYN 1966. *Field Crops of India.* 6th Edition. Bangalore Printing and Publishing Co. Ltd., Bangalore. pp 564.

Anonymous 1963. *Annual Progress Report for 1962-63.* Central Arecanut Research Station, Vittal, India. pp 117.

Anonymous 1964. *Annual Progress Report for 1963-64.* Central Arecanut Research Station, Vittal, India. pp 110.

Anonymous 1964. Warning against the spindle bug (*Carvalhoia arecae*) a serious pest of the areca nut palm. *Arecanut. J.* **15**: 22-23.

Anonymous 1967. *Annual Report of the Central and Regional Arecanut Research Stations, 1964-65.* Central Arecanut Research Station, Vittal. pp 92.

Anonymous 1971. Annual Report of the Central and Regional Arecanut Research Station, 1969. Central Plantation Crops Research Institute, Regional Station, Vittal. pp 79.

Anonymous 1983. *Annual Report for 1981.* Central Plantation Crops Research Institute, Kasaragod, 267 p.

Anonymous 2007. Research highlights. *CPCRI Newsletter*, **25** (4): 2-3.

Balasimha D and Rajagopal V 2004. *Arecanut.* Central Plantation Crops Research Institute, Kasaragod, 306p.

Bavappa KVA 1970. Mother palm selection in arecanut cultivation. *Indian farming.*, **20**: 31.

Bavappa KVA and Abraham KJ 1961. Influence of seed weight on the quality of seedlings in arecanut. *Arecanut J.*, **12**: 129-135.

Bavappa KVA, Kailasam C, Abdul Khader KB, Biddappa CC, Khan HH, Kasturi Bai KV, Ramadasan A, Sundararaju P, Bopaiah BM, Thomas GV, Misra LP, Balasirnha D, Bhat NT and Shama Bhat K 1986. Coconut and arecanut based high density multispecies cropping systems. J. *Plantn. Crops*, **14**: 74-87.

Bavappa KVA and Mathew T 1960. Effect of different spacings on quality of seedlings. *Arecanut J.*, **11**:3-8.

Bavappa KVA and Rarnachander PR 1967. Improvement of arecanut palm, *Areca catechu* Linn. *Indian J. Genet.*, **27**: 93-100.

Bhat KS and Abdul Khader KB 1982. Crop Management. B. Agronomy. *In: The Arecanut Palm.* (Eds. Bavappa KVA, Nair MK and Prem Kumar T). CPCRI, Kasaragod. pp 105-131.

Bhat NT and Mohapatra AR 1989. Effect of supplying nutrients through organic manures, inorganic fertilisers and their combination on arecanut crop. *J. Plantn. Crops*. **16** (Suppl): 443-447.

Chempakam B1998. Post harvest technology and product utilization in arecanut. *Harvest and Post harvest technology of plantation crops* (eds. Bosco SJD, Sairam CV, Muralidharan K and Amarnath CH) Central Plantation Crops Research Institute. Kasaragod, pp 74-78.

Dastagir AA 1965. A preliminary note on the new yellow leaf disease of areca palms in Mysore State ('Arasina roga' or 'chandi roga'). *Lal Baugh*, **10**: 3-4.

Dinesh Kumar EV and Mukundan K 1996. Economics of arecanut cultivation in Kerala. *J. Plantn. Crops*. **24** (Suppl.): 827-831.

Girish B, Shahapurmath H, Shivanna and Girisha HV 2003. Performance of Arecanut based mixed cropping systems. *Karnataka J. Agric. Sci.*,**16** (2): 254-259.

Kantha S, Ray BK and Lal R 1963. Laboratory evaluation of the toxicity of insecticides to palm mite (*Raoiella indica* Hirst.). *Indian Coconut J.*, **16**: 63-66.

Kumar R 2013. Threats and prospect of arecanut. *An economic study*, 14 pp.

Mohapatra AR and Bhat NT 1982. Crop management. A. Soils and manures. *In: The Arecanut palm.* (Eds. Bavappa KVA, Nair MK and Prem Kumar T). CPCRI, Kasaragod. pp 97-104.

Patel GI and Rao KSN 1958. Important diseases and pests of arecanut and their control. *Arecanut J.*, **9**: 89-96.

Peter KV 2002. *Plantation Crops*. National Book Trust, India, New Delhi, pp. 1-28.

Reddy KM, Saraswathy N. and Chandramohanan R 1978. Diseases of Arecanut in India- A review and further considerations. *Journal of Plantation Crops* 6: 28-4.

Sannamarappa M and Muralidharan A 1982. Multiple Cropping. *In: The Arecanut Palm* (Bavappa KVA, Nair MK and Premkumar T Eds). CPCRI, Kasaragod. pp. 133-149.

Shama Bhat K 1974. Intensified inter / mixed cropping in areca garden – the need of the day. *Arecanut & Spices Bull.* **5**: 67-69.

Sujatha S, Balasimha D and Bhat Ravi 2002. Fertigation of arecanut (*Areca catechu* L.) during pre-bearing stage. *Plantation Crops Research and Development in the New Millennium. Proc. PLACROSYM XIV*, 12-15 Dec. 2000. pp 328-332.

Thomas KG 1978. Crop diversification in arecanut gardens. *Indian Arecanut Spices &Cocoa J.* **1**: 97.99.

Venkatasubban KR 1945. Cytological studies in Palmae. Part I. Chromosome number in a few species of palms in British India and Ceylon. *Proc. Indian Acad. Sci., B*. 33:193-207.

Yadukumar N, Abdul Khader KB and Bhat KS 1985. Scheduling irrigation for arecanut with pan evaporation. Proc. Silver Jubilee Symposium on Arecanut Research and Development. CPCRI, RS, Vittal. Dec. 13-14,1982, pp 33-37.

OUTCOME ASSESSMENTS

PART A

Answer the following questions true or false.

1. Arecanut is commonly known as *supari*. True/False
2. Arecanut can withstand drought for a long time. True/False
3. Nut splitting is more a physiological disorder than a pathological problem. True/False
4. *Pueraria javanica* and *Mimosa invisa* are not good green manure-cum-cover crops in arecanut gardens. True/False
5. Arecanut palm is very delicate. True/False

PART B

Answer the following questions.

1. What is *Chali*?
2. Differentiate between *chali* and *kalipak.*
3. VTLAH 1 a hybrid between — and —.
4. How can you control fruit rot?
5. Good seedlings of 12-18/10-12 months old are selected for planting in the main field.

PART C

Write a brief note on each of the following.

1. Intercropping in arecanut.
2. Green manuring and cover cropping in arecanut.
3. Planting in arecanut.
4. Water management in betel palm.
5. Nutrition management in arecanut.

4

Production Technology of Oil Palm

Botanical Name: *Elaeis guineensis* Jacq.
Family: Arecaceae / Palmae
Chromosome No: 32
Place of Origin: Guinea Coast of West Africa
Common Name: Oil Palm

4.1 Introduction

The cultivated oil palm *Elaeis guineensis* Jacq. is a native of West Africa and exists in wild, semi- wild and cultivated forms. It was introduced as an ornamental crop in 1860 and planted at National Botanical Garden, Kolkata. Research on oil palm in India started during 1960 with the establishment of a research station at Thodupuzha by the Department of Agriculture, Kerala with Dura and Tenera germplasm imported from Malaysia and Nigeria. National Research Centre (NRC) for oil palm is situated at Elur, Andhra Pradesh. Oil palm crop is one of the highest oil yielding crops among all perennial crops. It gives 4 to 6 tonnes of edible oil/ha from 3 to 25 years of its life span compared to less than one tonne of oil per ha from other cultivated oil yielding crops. This oil palm is considered as golden palm due to its high yielding capacity. Among the 10 major oilseeds, oil palm is accounted for 5.5% of global land use for cultivation, but produced 32.0% of global oils and fats output. Palm oil shares 40% of the world's trade in edible oils (Oil World 2013). Almost 85% of the world's palm oil production is used as food. Oil palm is a crop of future and source for diversification, value addition, health and nutrition, waste land utilization and eco-friendly. It is considered to have tolerance to biotic and abiotic stresses. Presently, Southeast Asia is the dominant region of production with Malaysia being the leading producer and exporter of palm oil. Malaysian Palm Oil Board (MPOB) now has the largest oil palm germplasm collection in the world and high yielding plantations (Okwuagwu *et al.* 2011). Indonesia and Malaysia have raised large scale plantations of oil palm during the last two decades. India is importing about 5 to 8 lakh tons of palm oil every year from Malaysia, Indonesia for Public Distribution System (PDS).

Two distinct oils are extracted from oil palm, oil which is extracted from the pulpy portion, i.e., mesocarp of the fruit (contains 45-55 % oil) of oil palm is known as palm oil and it is edible, whereas the oil extracted from the kernel (50 % oil) is known as palm kernel oil which is used for cosmetics. Palm oil rich in carotenoids (major component) from which it derives its deep red colour oil is therefore promoted both as a 'balanced' oil (having equal saturated and non-saturated content) and one that is trans-free (Ascherio 2002).

In India, oil palm crop provides the excellent substitute of importing the oil. In India, oil palm is cultivated in more than 15 states by covering about 50,000 hectares under irrigated conditions and this crop is also cultivated under rainfed conditions. Major oil palm producing states in India are Andhra Pradesh, Karnataka, Assam, Kerala, Gujarat, Goa, Tamil Nadu, Maharashtra, Tripura, West Bengal and some areas of Andaman. Indonesia and Malaysia are the two major palm oil producers globally, producing nearly 85% of the global output. Palm oil substitutes import of edible oil. Palm oil is generally the cheapest commodity vegetable oil and also the cheapest oil to produce and refine globally. Oil palm is among the most productive and profitable of tropical crops for biofuel production. In 1992, the Oil Palm Development Programme (OPDP) was launched. This was followed by an "Oil Palm Area Expansion" (OPAE) programme in 2011-12. The government also allowed 100% FDI in palm oil plantations. In India over 154 exotic germplasm have been introduced from various countries by NBPGR, New Delhi.

Palm oil is consumed worldwide as cooking oil. Per 100 g of oil palm fruit contains 746 calories, 2.2% protein, 81.9% fat, 14.6% carbohydrates, 3.8% fat, 1.3% ash, 5.6% mg Fe, 50,680.6 μg carotene, 0.35 mg thiamine, 1.81 mg niacin, 136.1 mg Ca, 61.1 mg P and 0 % water. Majority of fatty acid in palm kernel oil is lauric acid (48%), myristic acid (16%) and oleic acid (15%). Palm kernel oil is about 82% saturated. Lauric acid is responsible for solid form of oil. Like other vegetable oil, it is also cholesterol free, having moderate level of saturation. This oil incorporated into fat blends is used in the manufacture of food products. Red or golden (refined) or red (unprocessed) palm oils are the major cooking oil in many parts of the world. This oil is rich in beta-carotene (precursor of vit. A), vit. E, tocopherols and tocotrieonols which are having antioxidant property. These natural antioxidants act as scavengers of damaging oxygen free radicals. Palm oil lowers the total blood cholesterol. Palm oil diet increases the production of hormone that prevents the blood clotting. Palm oil-based fats and margarine produce the most successful results in the preparation of bread and pies, biscuits, shortcakes and pastries. Among the non-food usages, palm oil and their products are used in soap making and oleo-chemicals which is used for preparation of environment friendly detergent, as they are biodegradable (Nampoothiri *et al.* 2006).

4.2 Soil and Climate

Oil palm can be grown in any type of soil, but best-suited soils are moist, well-drained, deep, loamy alluvial soils, rich in organic matter with good water permeability. At least one-metre depth of soil is required. Avoid highly alkaline, highly saline, waterlogged and coastal sandy soils. The oil palm tree is a tropical plant which commonly grows in warm humid climates at altitudes of less than 1, 600 feet above sea level with a mean annual temperature of 20-27°C, and temperature of more than 33°C inhibits photosynthesis. It is a sun loving plant, requires bright sunshine of more than 5 hours. Solar radiation below 350 cal /cm^2 /day affects the growth and yield of the palm. It requires evenly distributed rainfall of 150 mm/month or 2500-4000 mm/ annum. Rainfall distribution in India is not even and adequate. Hence grow oil palm under assured irrigation conditions by adopting recommended practices. Humidity of more than 80% is required to come up well.

4.3 Species and Cultivars

Cold and drought tolerant materials were collected from African countries namely Guinea Bissau, Cameroon, Zambia and Tanzania by CPCRI (Central Plantation Crops Research Institute) in collaboration with FAO in 1993. During the year 2000, few advanced generations of *Dura* accessions were imported from Costa Rica. These are given below:

Table 4.1: Future prospects for introduction

Sl.No.	Name of species	Desired traits	Source country
1	*Elaeis guineensis* Jacq.	Dwarf & high yield, high oil extraction ratio, good oil quality, compact canopy, drought tolerance and high FFB yield.	South East Asia, and Africa
2	*Elaeis oleifera* Jacq.	Dwarf & compact canopy, good oil quality	South America
3	*Elaeis guineensis* Jacq.	Dwarf & High yield	Central America
4	*Elaeis oleifera* Jacq.	Drought tolerance, compact canopy & quality oil	Central America

(Pedapati *et al.* 2013)

Varieties in oil palm

There are three different types of oil palm namely dura, pisifera and tenera based on their fruit forms. The significant differences among the three types are the presence or absence of shell and the thickness of shell.

1) Dura: It has low to medium mesocarp and it contains a thick shell around the kernel. It is not preferred for commercial cultivation. Mesocarp thickness varies in between 35 to 55% (54% by weight), shell thickness is 2-8 mm (30% by weight) and kernel 16% by weight.

2) Pisifera: It is a shell less fruit and pea like kernel inside. Embryo abortion is common in this variety and often kernel is also absent. The presence or absence of shell is genetically controlled. Dura is a genetic constitution of Sh+ Sh+ while Pisifera is Sh- Sh- and hybrid is Sh+Sh-. On selfing or inter-crossing, the hybrid fruits forms segregate into 25% Dura, 50% Tenera and 25% Pisifera.

3) Tenera: It is a hybrid between thick-shelled Dura and shell less Pisifera (F_1 hybrid). It is commercially/widely cultivated all over the world due to high proportion of mesocarp. It is characterized by the presence of thin shell and presence of distinct ring of fibres embedded in the mesocarp near to and encircling the seed and high oil content. Tenera fruits have a lot of pulp, thin shell with a big kernel. Thickness of mesocarp is 60 to 95% (74% by weight), shell is 10% by weight and kernel is 16% by weight.

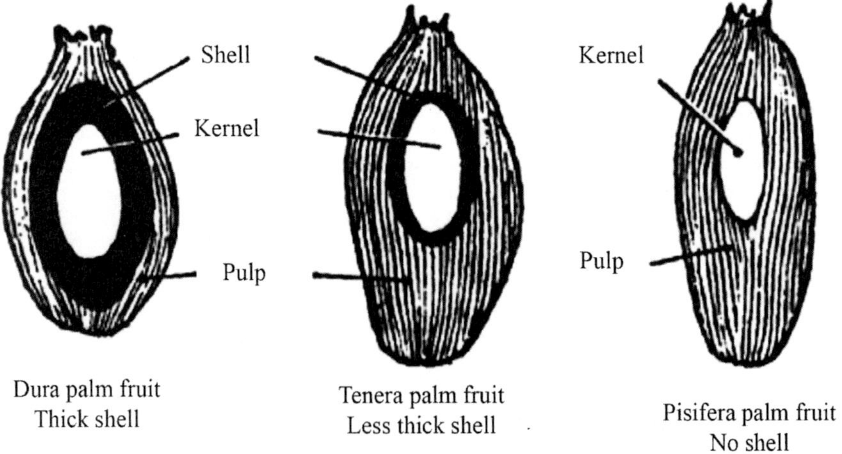

Fig. 4.1: Different parts of oil palm fruit

Table 4.2: Characteristics of different variety of oil palm

Sl.No.	Characters/Composition	Dura	Tenera	Pisifera
1	Mesocarp proportion in fruit (%)	35-50	60-96	98
2	Shell thickness (mm)	2 to 8	0.5 to 4	-
3	Oil percentage	15%	36%	25%
4	Average proportion of shell in fruit (%)	30	10	-
5	Average proportion of kernel in fruit (%)	16	16	10

Soh *et al.* (2003) have described the following breeding populations used in major breeding programmes:

1. Deli: A thick shelled dura cultivar from Bogor palms in Java. This provides the mother palms for almost all major commercial hybrid seed production.
2. Avros: This is a pisifera cultivar obtained from the seed from Eala Botanical Garden in Democratic Republic of Congo. These are precocious bearer, thin shelled, thick mesocarp and high oil content.
3. Yangambi: This is also a pisifera cultivar that produces bigger fruits and high oil content.
4. Ekona: It is a pisifera cultivar which has high bunch yield, good oil content and wilt resistance.
5. Calabar: It is a pisifera cultivar and used in seed production programme.

The ASD (Agricultural Services and Development) has collected many oil palm germplasm from different parts of the world and started breeding programme, out of which six high yielding cultivars were developed for different agro-climatic conditions. The cultivars produced by ASD are in commercial plantation. The first cultivar developed by ASD is Deli × AVROS. This variety produced large size bunches (> 15 kg), large fruit (> 11 g), excellent oil content (26-28%).

Deli × Ekona: Trees are moderately vigorous growing, medium bunch (13-15 kg), small fruit (< 9 g), high oil content (>28%).

Deli × Ghana: It produces medium bunch (13-15 kg), medium fruit (9-11 g), and high oil content (> 28%). They are moderately tolerant to drought, low temperature and low sunlight level.

Deli × La Me: Plant produces small size bunches (< 13 kg), small fruit (< 9 g), good oil content (> 26%). They have high tolerance to drought and moderately tolerant to low temperature and low sun light level.

Deli × Yangambi: Tree is vigorous growing, produces medium bunch (13-15 kg), large fruit (> 11g), high oil content (> 26-28%). They are highly tolerant to drought, low tolerant to low temperature and moderately tolerant to low sunlight level.

Three other hybrids, i.e., Tanzania × AVROS, Tanzania × Ekona and Bamenda × AVROS are cold tolerant.

4.4 Propagation

4.4.1 Nursery Management

Oil palm is propagated by seed. Well matured bunches (5 to 6 months old) are harvested for seed purpose. The bunch is cut into spikelets and fruits are taken out by hand. Then seeds are extracted from fruit. These seeds are then dried under shade for two days. Insufficiently dried seed germinates in storage and over dried seed (moisture content below 14%) leads to loss of viability. Properly dried seeds can be stored for a year at ambient temperature in covered drums or in polythene bags.

At the time of harvest, seeds are in dormant condition. Under natural conditions it takes about two years for germination. Soaking of seeds in hot water (50-60 °C) for 2 hours and treating the seed with Rhizobium culture @ 1 g per kg of seed improves germination percentage or soak the seeds in water for 5 days followed by drying in shade for 24 hours. The dried seeds are put in polythene bag and subjected to heat treatment in boxes of fermenting vegetable matter or in a germinator maintained at 40°C. After 80 days the seeds are taken out and soaked in water for 5 days followed by drying for two hours. The seeds are then put back to polythene bags and kept in a cool place to maintain the moisture content. Germination comes after 3 days and continues up to 30-40 days. In this method around 90-95 % germination occurs.

4.4.2 Nursery Raising

Polybags of 400-500 gauge and 40 cm × 35 cm size are used. The bags are filled with potting media and seed is sown in it. Good mulch is desirable during summer. Watering the seedling thrice in a week is recommended. A fertilizer mixer containing 15 g N, 15 g P and 6 g K @ of 8 g/ 5 lt of water for 100 seedlings may be applied at 2nd and 8th month stage. The seedlings are ready for transplanting in 12-14 moths (Peter 2002).

In Pisifera seed propagation is not possible as many of the fruits do not have embryo. Embryo abortion is common in this variety and often kernel is also absent. It is used as male parent in the production of Tenera. Pisifera palms are generally recovered from segregating populations since direct reproduction of this type is difficult due to the scarcity of fruits with embryo and the absence of protective shell.

4.5 Planting Systems and Method

Generally square and triangular systems of planting are used but the most commercially used method is triangular system. Planting should be done in pit size of 60 cm × 60 cm × 60 cm. Pit is to be filled up with 400 g SSP and 50 g phorate mixed with soil before planting. While planting, polythene bag should be taken off; the seedlings are to be planted with the soil, irrigation must be given right after planting. There are 143 numbers of plants per hectare that could be accommodated with a spacing of 9 m × 9 m × 9 m (triangular planting).

Best season for planting is June-December, i.e., during monsoon. In case of planting during summer, adequate irrigation, mulching and growing cover crops like sunhemp in the basin would help in avoiding hot winds during summer. Healthy seedlings of 12 -14 months old with 1-1.3 m height, 20-25 cm girth at collar region and 13 functional leaves are recommended for planting.

4.6 Gap Filling

Field inspection is to be carried out one to two months after planting to gap fill dead plants. They are to be given special care so that they can catch up with rest of the plantation. Yeow *et al.* (1981) reported early production of more female inflorescences in the initial 30 months is an indication of high yielders and all those that fail to produce female bunches within this period will remain as poor yielders. However, replacements may be affected to some extent by the vigorous growth of the neighbouring palms, which will shade the replanted palms. Renard and Quillec (1986) suggested to remove palms sensitive to *Fusarium* wilt at an early stage (2-5 years old plant) and to cover the soil with *Brachiaria* sp. instead of *Peuraria* sp. and carry out gap filling later.

4.7 Mulching

Mulching of polybags seedling is important aspect for nursery raising. Mulching with bunch refuse, saw dust, grass, wood shavings or kernel shells for at least 6 months during the early or late nursery period is beneficial to seedling growth. When mulching material is not short supply, sprouted seeds in polybags may be mulched in the early growing season (mid-April to November) to ensure that seedling will make adequate growth before the dry season or in the dry season only to ensure that final seedling vigour will not be reduced.

4.8 Effect of Shade / Light

From a study, Subronto and Taniputra (1979) observed that seedling raised under 61% and 79% light intensity were in most vigorous growth. They had highest leaf area, leaf number, rachis length, leaflet per rachis and bole girth. High light intensity at the nursery stage stimulated early flowering and resulted in a higher number of inflorescences per palm. Subronto *et al.* (1987) reported that in the field, there was no influence of nursery shade on the crop growth rate, net assimilation rate and vegetative dry matter accumulation.

4.9 Water and Weed Management

Oil palm requires sufficient irrigation, as it is a fast-growing crop with high productivity and biomass production. Do not grow oil palm if assured and adequate irrigation facility is not available. For grown up yielding palms of 3 years age and above, a minimum of 150 to 200 litres of water per day is required. However, in older plantations during hot summer this quantity may be increased up to 300 lit.

Basin method of irrigation is to be taken up when irrigation water is not a constraint. Required quantity of water is to be given at 4-5 days interval. Prepare irrigation channels in such a way that the individual palms are connected separately by sub-channel. For light soils, frequent irrigation with less water is to be given. In heavy soils, irrigation interval can be longer.

Drip or Microjet irrigation method is also practiced. If land is of undulated terrain, drip or micro sprinkler irrigation can be advantageous. If drip irrigation is installed, four drippers are to be placed for each palm. If each dripper discharges 8 litres of water per hour, 5 hr. of irrigation per day is sufficient to discharge 160 lit/day. In case of micro-sprinklers (180° or 360°) one each on either side of the palm can be installed. Drippers/jets should be periodically checked for proper discharge. Basins should be adequately mulched and covered with soil, which will help to conserve moisture. Yielding palm of 3 years age and above minimum of 150 litres/day is required.

Weeding

Take up regular weeding manually or with the use of only recommended herbicides. Use preferably contact herbicides, i.e., paraquat. Glyphosate (750 ml/ha/year or 17.5 ml/basin) is recommended for effective weed control. Herbicide mixtures of Paraquat with Atrazine, Monuron and Diuron sprayed on ground, twice a year can control the weeds, effectively.

Inter-cropping

Oil palm is a wide spaced perennial crop with a long juvenile period of 3 years. Inter- and intra-row space can be used to generate income during the juvenile phase of the crop. Intercrop selected should be compatible with the main crop and should not compete with oil palm for light, water and nutrients. Any remunerative crop can be grown, but the most suitable crops are vegetables, banana, flowers, tobacco, chillies, turmeric, ginger, pineapple etc. While growing intercrops in mature oil palm gardens of 8-12 years age or palms attained a height of 3 metres, intercrops should be able to grow under partially shaded conditions and should not compete with oil palm for water, sunlight and nutrients (e.g., cocoa, pepper, heliconia and ginger, lilly). Cover crops like *Pueraria phaseoloides, Calopogonium mucunoids, Centrosema prutascens, Mimosa invisa, Mucana* sp. can be grown.

4.10 Training, Pruning and Handling

4.10.1 Leaf Pruning

Development of leaves in the crown of palm is initially slow. Each leaf remains enclosed for about 2 years and then develops into a central spear (spindle leaf) before opening. The leaf stalk is strong and fibrous and is almost 8 m long. A mature leaf may have 250-300 leaflets; each about 1.3 m long and 6 cm broad. Rate of leaf production in oil palm is 20 to 25 leaves per year. Each leaf will also carry one inflorescence. The leaf bases adhere to the stem for about 12 years and longer and fall away gradually. Frond pruning in oil palm has influence on yield and hence is of economic importance. Severe pruning will adversely affect both growth and yield of palm, cause abortion of female flowers and also reduce the size of the leaves. Removal of senescent and useless fronds which are lying very close to the soil surface should be done in pre-bearing plant. It is suggested that the palms aged 4 -7 years should retain 6-7 leaves per spiral (48-56 leaves), those aged 8 -14 years 5-6 leaves per spiral (40-49 leaves) and those above 15 years should have 4-5 leaves per spiral (32 – 40 leaves).

4.10.2 Ablation

The bunches produced initially will be very small and have low oil content. Removal of such young male and female inflorescences and bunches during the first three years of oil palm growth is known as ablation. Removal of all inflorescences during the initial three years is found to improve vegetative growth of young palms so that regular harvesting can commence after three and half years of planting. Ablation is done at monthly interval by pulling out the young inflorescence using gloves or with the help of devices such as narrow bladed chisels. Ablation improves drought resistance capacity of young palms by improving shoot and root growth especially in low production areas where dry condition exists.

4.11 Nutrition

Oil palm is a gross feeder and demands a balanced and adequate supply of macro, secondary and micronutrients for growth and yield. It is advised to apply fertilizers at every three months interval.

Age of Oil Palm	Nitrogen	Phosphorus	Potassium	Magnesium	Boron
	Urea (g/palm/yr)	SSP (g/palm/yr)	MOP (g/palm/yr)	Magnesium Sulphate (g/palm/yr)	Borax (g/palm/yr)
1stYr	870	1250	670	125	25
2ndYr	1740	2500	1340	250	50
3rdYr onwards	2610	3750	2000	500	100

Four equal split doses of fertilizers are to be applied starting from June/July at three-month interval. For the newly planted crop, the first dose of fertilizer needs to be applied three months after planting. Add 50-100 kg FYM or 100 kg green manure per palm along with the second dose of fertilizer application, 5 kg neem cake/palm can also be applied. Broadcast the fertilizers around the clean-weeded basin, about 50 cm away from the palm base and incorporate into the soil with the help of fork. Irrigate the palms immediately after fertilizer application.

4.11.1 Micronutrients

Micronutrient elements, iron, manganese, copper and zinc are not generally found limiting in the nutrition of oil palm on acid soil conditions. Boron deficiency is occasionally found on young palms in the field showing a reduction of leaf area in certain leaves producing incipient 'little leaf', advanced 'little leaf' with extreme reduction of leaf area and bunching and reduction in the number of leaflets and 'fish-bone' leaf. The 'fish-bone' leaves are abnormally stiff with leaflets reduced to projections. Leaf malformations including 'hook leaf' and corrugated leaflets are some other associated symptoms. Soil applications of 50 - 200 g borax dehydrate, per palm, depending on age, and severity of symptoms is practiced for correcting the malady. Supply Mg @ 500 g per palm per year if deficiency symptoms are noticed.

4.11.2 Basin Management

During first year, basins of 1m radius, in second year 2 m radius, and in the third year 3 m radius are to be taken around the palm by removing the soil from inside so that the soil will not accumulate at the collar region. Basin area of oil palm represents its active root zone. Hence it must be kept clean and weed free to avoid competition for nutrients and water.

4.12 Flowering

Oil palm comes to flowering 14-18 months after planting. It produces both male and female flowers separately on the same palm. Male and female phases do occur naturally in consequent cycles in a palm.

4.12.1 Pollination

Oil palm is a highly cross-pollinated crop and pollination occurs either by wind or by insects, but wind pollination is not adequate. Effective pollinating insects like *Elaeidobius kamerunicus* helps in good pollination and fruit set. Release of this weevil after 2-1/2 year of planting is advisable. If the plants are not having good girth and vigour, release the weevils after 3 years. Introduction of weevil in India increased the fruitlet from 36.8 to 56.1% resulting in 40% increase in F/B ratio (fruit to bunch ratio). The maximum attainable pollination potential is as much as to cent percent with 57% increase in FFB weight.

4.12.2 Assisted Pollination

It refers to hand pollination for improving fruit set percentage. Assisted pollination is necessary to ensure fertilization of all female flowers in a bunch, especially during rainy season when wind pollination is not effective and numbers of palms in a plantation are on female phase. The inflorescence at anthesis is dried in the sun for about five hours. Then these are stored in moisture free container. It remains viable for a month. These pollens are mixed with chalk powder in a 1:14 ratio when used for pollination. The pollens can be used without mixing with chalk powder by dusting them on respective stigma. It is convenient to disperse the pollen on inflorescence by means of puffer. A lance puffer is used for taller palms (Peter 2002).

Sex ratio in oil palm is defined as the number of female inflorescences over the total number of inflorescences produced for a given period.

$$\text{sex ratio} = \frac{\text{No of female inflorescences}}{\text{Total no. of inflorescences}\left(\text{i.e., male +female}\right)} \times 100$$

4.13 Growth Regulators

Gibberellic acid is shown to increase the production of male inflorescences and reduce production of females in the oil palm. Ethephon and α-naphthalene acetic acid also have effects on sex ratio, but it is suggested that these may be indirect effects, in contrast to that of gibberellic acid which is considered to be a direct effect on the process of sex differentiation. Pre-harvest application of ethylene releasing agent ethephon (2-chloroethyl phosphonic acid) to attach bunches of oil palm fruit, increases oil content by 7%. There is a relationship between increased ethylene levels and increased palm oil content.

4.14 Crop Protection

4.14.1 Insect-Pests

1. Tussock Moth (*Dasychira mendosa* Hb.) (Lepidoptera: Lymantridae)

It is a polyphagous insect, which is generally found on arecanut, citrus, guava, banana etc. Larvae occasionally cause significant damage to nursery seedlings by feeding voraciously on the leaves. They are found causing damage in oil palm plantations also (Wood 1968). Occurrence of this pest was noticed in oil palm nurseries of Kerala, Karnataka, Andhra Pradesh and Tamil Nadu. The pest is noticed throughout the year with the highest incidence during June - July coinciding with the onset of heavy rains (Dhileepan 1992). Incidence is seen only in those nurseries which are over grown and not lifted for transplantation at correct time. The larvae are found defoliating the oil palm severely at secondary nursery stage. Initially the young larvae scrap the leaves in congregation and disperse in later stages and start defoliating the tender leaves severely.

Management: Under field conditions, the larvae are parasitized by tachnid flies to the extent of 10% and pupa are parasitized by *Brachymeria albotibialis* to the tune of 40.0%. The pest can be managed with one or two sprays of quinalphos @ 0.05%.

2. Rhinoceros Beetle (*Oryctes rhinoceros* L) (Coleoptera: Scarabaeidae)

It is a common pest of many palms including coconut, oil palm, areca nut and palmyrah palms in all the states of India. Pest incidence is found throughout the year. Peak period of adult emergence is during South-West Monsoon (June to September). Infestation is severe in plantations where field hygiene and sanitation are neglected. Incidence is more in oil palm plantations adjoining coconut gardens.

Damage: Adult beetles bore into the palms at the base of the spear cluster to consume the sap and tender parts of the leaves when it is not opened. Through the outermost petiole of the spear cluster, the beetles penetrate to the interior, leaving it permanently marked with a hole. The wedge-shaped gap in the leaf silhouette and a hole in the petiole are the common characteristic symptoms. Young palms exhibit much more severe damage at the base of the spears as compared to mature palms. The damaged spindle may collapse or expanded fronds may snap off or be truncated. Adult rhinoceros beetles are found boring and chewing the male and female inflorescences even before anthesis, while the inflorescences are inside the spathe. The entry holes of beetles can be recognized by the presence of chewed up fibrous tissues. Secondary rotting to the bud is commonly seen due to the entry of fungi and bacteria through the injuries made to the heart of the palm by the pest. The injuries made on the petioles and female inflorescences serve as sites for egg laying of red palm weevil. Presence of more lignin at the place of feeding (leaf petiole) in oil palm compared to leaf tips in case of other arecaceae palms is the prime reason for low pest incidence of the pest (Kalidas 2011).

Management: The beetles, which burrow deep into the crowns of young palms, are to be extracted by means of a hooked pointed metal rod (beetle hook). After extraction of the beetle, the leaf axils around the injured spindle/leaf are to be filled with mixture of Mancozeb and sterilized fine sand at a ratio of 3 g: 1 kg. The *Baculo* virus of *Oryctes* is one of the most successful microbial control agents employed for the bio-suppression of rhinoceros beetle infesting coconut. Release of the infected beetles is the most economical, effective and easy method for dissemination of the viral inoculum into the natural population of the beetles. For seedlings, apply 3 naphthalene balls/palm weighing 3.5 g each at the base of inter space in leaf sheath in the 3 inner-most leaves of the crown once in 45 days. Apply mixture of either neem seed powder + sand (1:2) @150 g per in the base of the 3 inner-most leaves in the crown.

3. Red Palm Weevil *(Rhynchophorus ferrugineus)*

The palms attacked by rhinoceros beetle are more prone to infection by red palm weevil.

Damage: The grub bore through and feed on softer tissue of stem, meristem and mesocarp of fruits. Being concealed in plant tissues, the pest can be detected only at a very late stage, often when the crown topples. The symptoms of the infested palms include gradual wilting and drying of outer whorl of fronds, presence of small round holes on the stem, longitudinal splitting of leaf base and gnawing sound produced by feeding grub inside the crown. In some cases, rotting of spear can also be noticed. The red palm weevil in most of the cases lays eggs on injuries made by rhinoceros beetle on bunches. The larvae feed on the mesocarp of fruits and the infection causes qualitative and quantitative loss of the fruits.

Management: Remove and burn all wilting or damaged palms in oil palm gardens to prevent further perpetuation of the pest. Avoid injuries on stems of palms as the wounds may serve as oviposition sites for the weevil. The wounds, if any, on the palm are to be treated with coal tar + insecticide to prevent entry of the pest to the palm trunk. Fill all holes in the stem with cement. Setting up of attractant traps (mud pots) containing sugarcane molasses 2½ kg or toddy 2½ litres + acetic acid 5 ml + yeast 5 g + longitudinally split tender oil palm stem/logs of green petiole of leaves of 30 numbers in one acre to trap adult red palm weevils in large numbers. Install pheromone trap @1/2 ha. Fill the crown and the axils of top most three leaves with a mixture of fine sand and neem seed powder or neem seed kernel powder (2:1) once in three months to prevent the attack of rhinoceros beetle damage in which the red palm weevil lays eggs. Trunk injection with carbaryl 50% WP at 1 % or endosulfan 35 EC at 0.1% is an effective curative treatment for palms with advanced stage of infection.

4. Slug Caterpillar *(Darna catenatus snellen)*

Damage: Caterpillars are found feeding on the leaf lamina causing heavy defoliation leaving only midribs. The incidence is very heavy on the lower whorl leaves making them completely dried. Intensity of the pest could be assessed based on the presence of faecal droppings on the ground level/cover and gnawing sound during severe outbreak.

Management: The lower fronds having pest stages are to be pruned at the beginning of the pest activity period and burn them. Application of microorganism, *Beauveria bassiana* (10^{-1}) that caused white muscardine disease to the caterpillar proved effective but time taking compared to chemical insecticides. Under natural conditions, large number of parasitoids of *Eulophidae, Chalcididae, Braconidae, Tachinidae*, and *Bombyliidae* and predators of Hemiptera and Pentatomidae families regulate the host populations (Mariau 1999). Spraying of carbaryl 50% WP at 0.1 % is recommended for the control. Aerial spraying of quinalphos @ 0.05% or lamdacyhalothrin @ 0.05% twice during pest activity period at 15 days interval can effectively control the pest. Stem injection with monocrotophos 25 ml/100 ml of water per palm is also effective when the palms attained height.

5. Scales and Mealy Bugs

Dysmicoccus brevipes infests the pre-anthesising male and female inflorescences and also unripe and ripe oil palm fruits. Mealy bugs are found in moist, warm climates. Whereas poor hygienic conditions/sanitation practices that are being followed in the gardens are the major criteria for endemic infestation.

Damage: Coccids (soft scales) are found on the leaves of oil palms of all ages. Diaspids (armoured scales) are commonly found on the fruit bunches and leaves of oil palm. Both nymphs and adult females suck sap from the tender spear leaves, inflorescence and fruits. Attack by Diaspids results in appearance of chlorotic spots on leaf tissues. These pests secrete a waxy cover, which hardens to form a tough armour. *Dysmicoccus* spp. infests the oil palm fruits of Fresh Fruit Bunches (FFB) by sucking the sap from mesocarp. These pests feed on plant sap and excrete honey dew which will attract ants and sooty mould development.

Management: Conservation and augmentation of natural enemies such as ladybird beetle or release of coccinellid beetle, *Cryptolaemus montrouzieri* @ 10/tree reduces mealy bug population. Ladybird beetles are the most important predators of mealy bugs. Infestation of mealy bugs can be controlled by spraying with dimethoate at 0.2% or methyl demeton at 0.025%. Scale insects can be controlled by spraying with Malathion 0.1%.

6. Avian Pests

Birds and rodents are the major pests of oil palm in the oil palm growing countries of the world. Both of them feed on the mesocarp of Fresh Fruit Bunches (FFB) leaving only fibers on the seeds and cause direct losses on yield. Damage by birds is either partial or complete. In partially damaged fruit, 40-50% total weight of individual fruits is eaten away by birds. In bunches with total fruit damage, weight loss of 68-73% can be observed. In many ripe fruit bunches, all the fruits are lost resulting 100% loss of fruit weight.

Management: The ripe fruit bunches after 150 days of fruit set are to be covered with wire net of 1.25 cm mesh (60 × 90 cm size), reed baskets, plaited coconut leaf baskets or oil palm leaves to avoid bird damage (Dhileepan and Jacob 1996). Covering the bunches with oil palm leaf tips and tying with a piece of rope to keep them firm and impenetrable by the bird beak is found to be effective and cheap. Tying nylon fishnets of 9 × 1 m size in between two palms is found to be most effective to manage the menace. Nets having 5 sq cm size holes are best fit to trap all the birds. An average of five nets per ha could give maximum benefit.

7. Rat

Burrowing rat, *Tatera indica* (Hardwicke) was found attacking the young oil palm plants. They burrow down to the bole region by making cavities to feed the cabbage tissue resulting into the wilting of leaves and mortality of the palms. Black rat, *Rattus rattus* Linn. mainly feed on the immature fruit bunches of 2-3 months old. They feed on the apical part of mesocarp and kernel portion of the fruit which is semi solid. The symptoms of attack depict as half cut of the fruits.

Management: Damage to young seedlings can be prevented by placing barriers consisting of 1.25 cm mesh (Chicken wire mesh) collars around their base. They must be tightened around the palm and well fastened down to prevent the rats getting inside or underneath the ground. Traps such as iron live traps, death fall trap, bow trap etc. may be used as an integrated approach to minimise the rodent damage to oil palm. Baiting with zinc phosphide and bromadiolone are found effective against rat menace.

8. Leaf Web Worm *(Acria meyricki)*

The infestation aggravated in those orchards where the palms were given basin as well as flood irrigation with excess quantity than the required. A yield loss of up to 34% is reported for leaf web worm. The incidence is generally observed from October to April.

Damage: The caterpillar stays inside a web on the underside of the leaves. Early instars scrap the leaves and later instars cause defoliation. Initially they are found feeding on the older leaves causing heavy defoliation. After complete defoliation of the lower leaves the caterpillars migrate to the next upper leaves. Due to severe infestation, the leaflets are dried and give burnt up appearance. Upon disturbance, the larvae hang on silken threads and spread to either nearby palms or retrieved back and feed on the same leaves. Intermingling of palm fronds in the garden paves way for easy spread of infestation.

Management: The lower fronds having pest stages are to be pruned at the beginning of the pest activity period and burn them. Under field conditions, larvae are found parasitized by two biocontrol agents *viz*., *Apanteles hyposidrae* and *Elasmus brevicornis* Gahan. Pest is also effectively controlled with the application of the microbial organisms, i.e., *Beauveria bassiana.* Aerial spraying of quinalphos @ 0.05% or lamdacyhalothrin @ 0.02% twice during pest activity period at 15 days interval is proved effective in controlling the pest. Stem injection with monocrotophos 25 ml/100 ml of water per palm is also effective in case of tall palms.

4.14.2 Diseases

1. Bacterial Buds Rot (*Erwinia* spp.)

Parts of spear leaf petiole or rachis turning brown; discolouration may be associated with a wet rot; spear leaf may be wilted and/or chlorotic; leaves may be collapsing and hanging from the crown; infection of the bud results in buds becoming rotten and putrid, leading to death of the palm.

Management: Plant oil palm cultivar with resistance to the bacteria; rotting tissue on spear leaves should be removed to prevent bacteria spreading to buds; palm buds can be protected using copper-based fungicides like Blitox 50.

2. Ganoderma Butt Rot (*Ganoderma* spp.)

Reduced growth of palm, pale green foliage, older fronds turning chlorotic or necrotic; drooping fronds; on mature oil palms, spear leaves do not open, seedlings may exhibit a one-sided chlorosis or necrosis of the lower fronds; cross sections of lower portion of trunk reveal a discolouration and softening of the central area and a distinct boundary is present between healthy and diseased tissue. It is caused by fungi.

Management: There are currently no fungicides recommended for protecting palms from Ganoderma butt rot; palms should be monitored closely for signs of disease, especially if a palm has died or been removed nearby as fungi can colonize old stumps and release spores; infected trees should be removed as once symptoms are present in foliage, a large portion of the trunk is already rotted and the palm is unstable; do not replant palm in soil where an infected palm has been removed.

3. Oil Palm Wilt (*Fusarium oxysporum*)

Symptoms of the disease vary with age of host; disease can affect seedlings and mature trees; seedlings exhibit retarded growth, reduced leaf size, chlorosis of older leaves and tip necrosis; field palms may exhibit a bright yellow chlorosis of leaves in the mid-canopy which starts at the tip of the pinnae and moves towards petioles before affecting adjacent fronds and spreading to older leaves in the canopy; in older palms, lower leaves wilt and dry out and fronds break close to the base of the trunk; new fronds are chlorotic and stunted; the palm shows decline on one side and develops symptoms in the lower canopy; infection spreads rapidly upwards and infects the bud, killing the palm. Fungus infests palms through the root system.

Management: International quarantine procedures have limited the spread of the disease between major palm oil producing countries; dead or dying trees should be felled and burned to prevent spread in plantations; if palms are replanted then new palm should be planted a distance of 3.9 m from infested stump; soil within a 3 m radius of infested stumps should be treated with dazomet and covered for a period of 30 days.

4. Leaf Spot (*Pestalotiopsis* spp.)

Tiny black spots on leaves which enlarge into 2 mm long elliptical, elongated lesions; lesions may expand and be surrounded by black tissue and chlorosis between lesions; lesions may be present on leaf petioles and rachis. This disease is caused by fungi.

Management: If palm is severely diseased, it should be removed from plantation and destroyed; palms should be planted with adequate spacing to allow air to circulate between trees; remove weeds from around palms; applications appropriate broad spectrum foliar fungicides can help to protect the palms from disease.

5. Bud Rot

Yellowing of spear leaves and bending of the affected spear at the base. Spear is seen hanging down among the healthy fronds. In young palms 2 to 3 leaves of the youngest whorl become necrotic and dry up. Disease incidence is severe during rainy season, when atmospheric temperature is low and relative humidity is high. Retention of disease affected palms in the field results in the inoculum build up and often the disease becomes rampant. Rotting starts at the basal portion or collar region of the spear close to the meristem. Rotting tissues appear pale to pink with a brown border and emit offensive odour. The basal tissue of the spear rots completely, collapses and comes out without much resistance when pulled out. Rotting advances downward and a big crater is formed in the centre of the crown affecting the apical bud tissues. If unchecked, rotting leads to the total destruction of meristem and death of the palm.

Management: Early detection of affected palms, cleaning of the rotten portion and drenching with Carbendazim 0.1% solution saved the palms. In the recovery phase, the palm produces short leaves with twisted petioles. In garden having severe disease incidence, prophylactic drenching of Carbendazim 0.1% solution to the spear clusters twice, before and after the monsoon protects the palms from bud rot incidence.

6. Bunch Rot (*Marasmius palmivorus*)

In estates with good harvesting standards and up-to-date pruning of old fronds, *Marasmius* bunch rot is not a problem. However, when rotten bunches are available for colonization and coupled with wet seasons of high humidity it can attack healthy tissue like developing fruit bunches and frond bases and gradually spread to younger bunches and inflorescence. Serious outbreaks have been recorded in plantations with inadequate pollination, poor sanitation in wet seasons, damaged/aborted bunches due to pest attack and acid sulphate soils with high incidence of bunch abortion (Turner and Gillbanks 2003).

Damage Symptoms: During the initial stage of infection, strands of mycelium are seen on top of the fruit bunches. When conditions are very moist or humid, fruit bunch is rapidly covered by many whitish mycelium and rhizomorphs. In the later stage, the mycelium grows over the fruits and penetrates the mesocarp to produce a wet rot in light brown colour. The fruit bunches finally rot and abort due to *Marasmius* infection and secondary infection by other micro-organisms. After some time, *Marasmius sporophores* in the form of whitish mushroom like fruiting bodies can be seen.

Management: In view of its widespread presence in the oil palm ecosystem, it is not possible to eradicate *Marasmius*. The disease can be controlled by keeping palm sanitation to remote rotten bunches, improving pollination and improving nutrition in acid sulphate soils. A solution containing 0.1 % Carbendazim + 0.1% Monocrotophos can be sprayed.

4.15 Physiological Disorders

Bunch Failure

Failure in the development of bunches at any stage during anthesis to harvest is referred as bunch failure. Periodical palm cleaning reduces the load of inoculums and fresh incidence.

Cause: Specific cause is not known; however, it may be due to excess pruning, mutual shading, under pollination, release of pollinating weevil, moisture stress/prolonged drought, inadequate nutrient status, over bearing etc. increases the rate of bunch failure.

Control Measures: There is no recovery once bunch failure has started and hence all control measures must be aimed at avoiding those conditions favouring bunch failure.

4.16 Harvesting

Pre-bearing age in oil palm is three years and economical life span is 25 years. Fruit is ready for harvesting when the fruit colour changes from purple to orange colour and the single fruit could easily fall off from the bunch. Fruit takes about 180 days (6 months) from pollination to maturity. Fully ripe fruit bunches should be harvested as immature bunches and partially rotten bunches are not suited because it results in low oil recovery of poor quality. Over ripe fruits reduce quantity and quality of oil. If harvesting is delayed, the fat is converted to free fatty acids and glycerol. As the oil synthesis and free fatty acids formation occur during ripening process, harvesting should be carried out as frequently as possible in order to reduce the number of over ripe bunches. Over ripe bunches have a high degree of fruit detachment and have increased oil acidity. In a fresh ripe, un-bruised fruit the fee fatty acid (FFA) content of the oil is below 0.3 %. However, in ripe fruits, the exocarp become soft and is more easily attacked by lipolytic enzymes, especially at the base when the fruit becomes detached from the bunch. The enzymatic attack results in increase in FFA of the oil through hydrolysis. Research has shown that if the fruit is bruised, the FFA in the damaged part of the fruit increases rapidly to 60% in an hour. According to current practices, harvesting should be done at every 7 to 14 days intervals.

Maturity indices

1) Colour change: When colour of fruits changes from black to orange or red or yellowish orange, fruit is ready for harvesting.
2) Detachment of fruits: For practical purposes when few fruits (around 10 fruits or more detached or easily removable from young palm and 5 fruits for adult palm) are detached from the bunch.
3) Change in fruit texture: Fruits become smooth when ripe and fruits can be pressed with fingers with ease.

Yield: The average weight of harvested fruit bunches is 30-40 kg and average number of bunches developed per palm nearly 10 to 12. In one hectare of area, average FFB yield is 12 tonnes and oil recovery out of FFB is 18 to 21%, i.e., extraction ratio from oil to bunch is 20 %. However, under good management practices, 20 to 30 tonnes of FFB is harvested from one hectare from which 4 to 6 tonnes of oil is obtained. However, on an average 12 tonnes of fresh fruit bunches (FFB) can be obtained per ha per year, yielding 2.5 tonnes of palm oil.

4.17 Post-Harvest Handling and Processing

Harvested fresh fruit bunches (FFB) have to be transported to the factory as quickly as possible and at any cost not more than one day (within 24 hours). The fruits of oil palm should be processed within few hours after harvest to obtain good quality oil. There will be deterioration of oil due to over ripening, storage, damage of fruits, etc. There are two types of oil extracted from fruit, one is edible from fruit mesocarp. It is known as palm oil and it contains around 45-55% oil and colour of oil varies from light yellow to orange red depending on the amount of carotenoids present in it. The second type oil is extracted from fruit kernel or endosperm. It contains 50% oil and it resembles coconut oil, i.e., nearly colourless.

As oil is extracted from mesocarp portion of fruit, the method of oil extraction is entirely different, i.e., wet processing. One of the major problems in oil extraction in oil palm is deterioration of oil into free fatty acids which results in poor quality of oil. FFA (free fatty acid) content should be less than 2% for using it as edible oil. During processing oil palm fresh fruit bunches are sterilized at 130ºC for one hour under pressure of 2 kg per cm^2 to inactivate the fat splitting enzymes. Sterilization process leads to inactivation of lipase and lipolytic enzyme activity i.e., fat splitting enzyme which are responsible for increase in FFA, loosen the fruit for easy separation, soften the fruits for facilitating digestion and coagulation of mesocarp proteins.

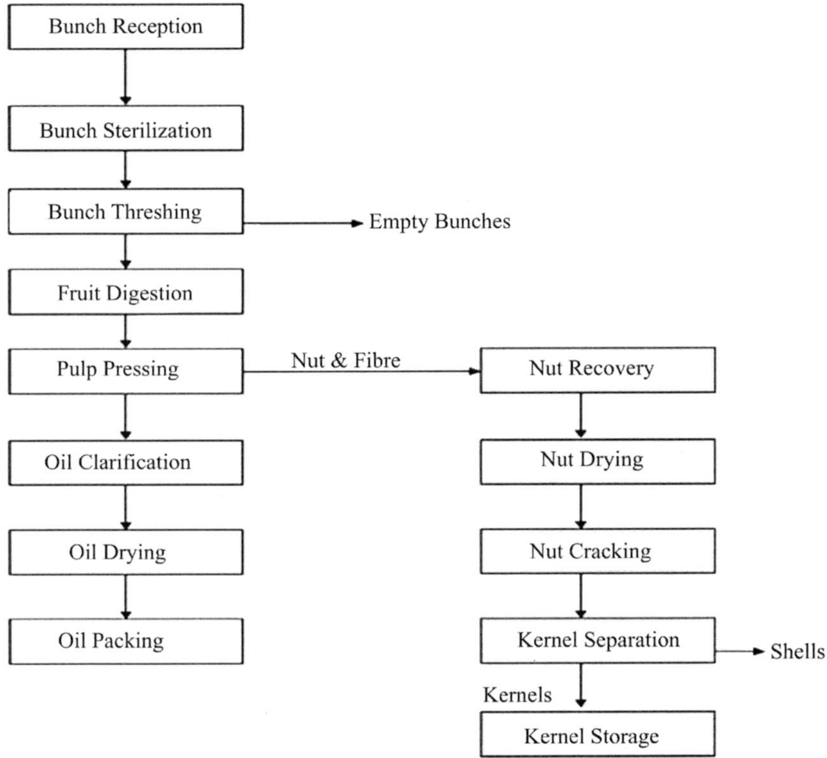

Fig. 4.2: Palm oil processing unit operations

In oil winning process, in summary, involves the reception of fresh fruit bunches from the plantations, sterilizing and threshing of bunches to free the palm fruit, mashing the fruit and pressing out the crude palm oil. The crude oil is further treated to purify and dry it for storage and export. Large scale plants, featuring all stages required to produce palm oil to international standards, are generally handling from 3 to 60 tonnes of FFB per hectare. The large installation has mechanical handling systems (bucket and screw conveyers, pumps and pipelines) and operates continuously, depending on the availability of FFB. Boilers, fuelled by fire and shell, produce superheated steam, used to generate electricity through turbine generators. The lower pressure steam from the turbine is used for heating purpose throughout the factory. Most processing is automatically controlled and routine sampling and analysis by process control laboratory ensure smooth and efficient operation.

4.18 Yield

Yield of oil from oil palm is 4 to 6 tonnes per ha compared to less than one tonne per ha from other cultivated oil seed crops (while other oil seed crops yield < 599 kg oil per ha (i.e., < 0.60 tonnes per ha)). If five lakh ha of area suitable for oil palm cultivation is developed under oil palm (Agroclimatic conditions), the demand and supply gap (Production and consumption gap) can be successfully met.

4.19 Economics

Raw material is supplied to industries like vanaspati industry, soap industry and also used in production of oleo-chemicals products such as fatty acids, fatty alcohols, gylcerols and other derivatives (these are used for the manufacture of cosmetics, pharmaceuticals, detergents and industrial products). Different by-products such as fibre from fronds and empty fruit bunches are used to make medium density fibre boards; furniture is produced from the trunk and mushroom is cultivated on empty bunch. The detailed cost of production of one tonne of palm oil is given below:

Table 4.1 Costs and returns from oil palm processing (Rs. / tonnes)

Sl.No.	Particulars	Amount	Percentages
I. Costs			
Variable Cost			
1	Cost of raw material	16176.47	85.80
2	Incidental charges	153.66	0.81
3	Wages for casual labour	380.88	2.02
4	Power charges	38.60	0.21
5	Lubricant charges	2.60	0.01
6	Fuel charges	57.90	0.81
7	Miscellaneous charges	30.94	0.16
8	Interest on working capital	5.69	0.08
	Total variable costs	16846.74	89.35
Fixed costs			
1	Depreciation on buildings	174.67	0.93
	i. Factory building	174.67	0.93
	ii. Administrative building	30.19	0.16
	iii. Work shops	1.48	0.01
2	Depreciation on		
	Machinery	327.95	1.74

	Effluent treatment plant	45.63	0.24
	Tractor and accessories	30.64	0.16
	Drums (205 lt)	5.11	0.03
	Fire extinguishers &Fixures	1.04	0.01
	Generator	4.19	0.02
3	Opportunity cost of land	29.14	0.16
4	Repairs and maintenance	57.44	0.30
5	Insurance charges	38.65	0.20
6	Taxes	74.71	0.40
7	Salaries for permanent staff	94.07	0.50
8	Interest on fixed capital	1092.37	5.79
	Total Fixed costs	2007.55	10.65
	Total cost	18854.29	100.00
II returns			
	Returns from one tonne of oil	36,000	92.45
	Returns from kernels	2940	7.55
	Gross Returns	38940	100.00
III	Net Returns from one tonne of oil	20086.01	

(*Source*: Srilatha, 2017)

References

Ascherio A 2002. Diet fat and risk of CHD: New developments. In: Enhancing Oil Palm Industry Development through environmentally technology, Z. Poeloengan *et al.* (Eds.) Indonesian Oil Palm Research Institute. pp. 60-65.

Corley RHV 1976. Sex Differentiation in Oil Palm: Effects of Growth Regulators. *Journal of Experimental Botany,* **27** (98): 553-558.

Dhileepan K and Jacob SA 1996. Pests. In: *Oil Palm Production Technology*. pp. 49-58. CPCRI, RC, Palode.

Kalidas P 2011. Strategies on pest management in oil palm. In: *Recent trends in Integrated Pest Management* (Eds. AK Dhawan, Balwinder Singh, Ramesh Singh and Manmeet Brar Bhuller). Indian Society for the Advancement of Insect Science. pp.177-185.

Nampoothiri KUK, Parthasarathy VA, Krishnakumar V and Ponnamma K 2006. Oil Palm. *Plantation Crops*. Vol-2, pp-283-420. Naya Udyog Publication.

Okwuagwu CO, Ataga CD, Okoye MN and Okolo EC 2011. Germplasm collection of highland palms of afikpo in eastern Nigeria. *Bayero J. Pure and Applied Sciences*. 4: 112-114.

Pedapati A, Tyagi V, Yadav SK and Murugesan P 2013. Present status and future priorities for introduction of oil palm in India. The Ecoscan. **7** (3&4): 139-144.

Peter KV 2002. Oil Palm. *Plantation Crops*. National Book Trust. Pp203-227.

Renard JL and Quillec G 1986. Oleagineux, **37**: 397-404.

Srilatha C 2017. Economic aspects of oil palm processing in Nellore district of Andhra Pradesh. *International Journal for Research in Emerging Science and Technology*, **4** (9): 9-12.

Subronto and Taniputra B 1979. *Bull. Bali Penelitian Perkebunan*, **10**:77-86.

Subronto, Taniputra B and Manurung A 1987. *Bull. Perkebunan*, **18**: 127-136.

Yeow KH, Tarn TK, Poon YC and Toh PY 1981. In: *The oil palm in Agriculture in the Eighties*. Inc. Soc. Planters, Kuala Lumpur.

OUTCOME ASSESSMENTS

PART A

Answer the following questions true or false.

1. Three distinct oils are extracted from oil palm. True/False
2. Yield of oil from oil palm is 4 to 6 tonnes per ha. True/False
3. Harvested fresh fruit bunches (FFB) have to be transported to the factory as quickly as possible. True/False
4. Pre-bearing age in oil palm is 10 years. True/False
5. Tenera fruits have a lot of pulp, thin shell with a big kernel. True/False

PART B

Answer the following questions.

1. Oil Palm Wilt is caused by —.
2. Oil palm is a self-/cross-pollinated crop.
3. What is bunch failure?
4. Yield of oil from oil palm is — to — tonnes per ha.
5. What is dura?

PART C

Write a brief note on each of the following.

1. Ablation.
2. Harvesting of oil palm.
3. Water and weed management in oil palm.
4. Nursery management in oil palm.
5. Assisted pollination.

5

Production Technology of Palmyrah Palm

Botanical Name: *Borassus flabellifer* L.
Family: Arecaceae
Chromosome No: 36
Place of Origin: Tropical region of Africa, Asia and New Guinea

5.1 Introduction

Palmyrah is a monocotyledon and perennial crop. The word "Borassus" was derived from a Greek word and it means the leathery covering of the fruit and "flabellifer", means fan-bearer. The distinguishing characters of palm in this genus are their palmate, fanlike leaves and dioecious character, i.e., male and female flowers are borne on separate trees (sex ratio 1:1), allotetraploid (X= 8 or 9, n = 18), the male tree has a heteromorphic XY pair of chromosomes while the female has an XX pair (Sankaralingam *et al.*, 1999). Next to coconut, palmyrah is the most abundant palm found in the world. Palmyrah thrives in arid conditions and is grown extensively in Tamil Nadu. Palakkad district of Kerala is popularly known as land of palmyra trees. Palmyra trees are known as the icon of this district and have a vast cultural, heritage and literary association.

Compositions and Uses

Palmyrah palm is basically a nutritionally fortified fruit. Its nutritional potential is illustrated in a tabular way below:

Elements	Concentration (mg/100g)	Amino acid	Concentration (g/100g protein)
Calcium (Ca)	8.76	Aspartic acid	2.6
Phosphorus (P)	15.00	Glutamic acid	0.41
Iron (Fe)	1.20	Alanine	0.7
Magnesium (Mg)	10.2	Proline	6.3
Zinc (Zn)	3.00	Phenyl alanine	0.41
Copper (Cu)	10.0	Lysine	0.8
Sodium (Na)	20.0	Tryptophan	3.3
Potassium (K)	21.5	Aspargine	2.6

Source: Mani and Dutta, 2018

The palmyrah offers its produce to the people throughout the year. There is Neera (the sweet juice of its fruit), Toddy (obtained by fermenting the sugary sap of the shoots), Nungu (a fruit with a soft inner kernel holding sweet water which has cooling properties), panangaai, Palm fruit, palmyra root, thavun, jaggery all of which have healthy nutrients in them. The leaves, husks, fibre, karukku, beam and pannadai can be used to make boxes, baskets, mats, brooms, ropes, fans, winnows and 20 other items which can be used by rural people for their daily chores. A large variety of handicraft can also be made not only for personal use but also for the market. Its leaves are made into brooms and also used for thatching. The Palmyra can also be used for fencing, making thatched dwelling, roofs for houses and providing beams. Palmyra is slow grower. First frond appears in about 5 months. First fan shaped tree leaves appear only in the second year.

Botany

Roots: Root system is adventitious in nature. Roots are originated from internode at the base of the stem. The main roots are string like giving rise to many lateral branches, which branch off further. Mature roots have sclerotic exodermis. The roots of the palm store water.

Stem and leaves: Crown of palmyrah palm contains 30-40 palmate leaves, which measure 1-1.5 m in diameter. One leaf per month is produced and divided into 60-80 segments, which measure about 3 cm broad at the base. The petiole is 0.6-1.2 m long, strong and having serrated margins. Stem is unbranched and topped by a crown of 20-30 large leaves. Each leaf has a stout long petiole of 0.9 to 1.5 m long and rachis with lamina. The petiole base is broad having a vertical split and it is persistent. It clasps the stem almost half the circumference and is surrounded by a fibrous network of stipules. Leaf lamina is palmate, large, 1.0 to 1.5 m long along with 60-80 compound segments. Foliar spiralling is either left or right-handed with alternate phyllotaxy, that helps efficient utilization of sun light. Cortex in the stem is narrow bearing vascular bundles that have massive phloem sheaths. Parenchyma cells are close to the vascular bundles and are isodiametric in shape and found in groups.

Inflorescence: Sex in palms can be differentiated only during flowering. Flowering is seasonal, February-May and fruit mature in August-September. The inflorescence is inter-foliar and branched spadix, sheathed by many imbricated, fibrous, coriaceous spathes. The outer most spathe is the smallest one while the inner most one is the longest. A palm produces 5-8 spadices annually.

Male spadix: The male spadix has 5-10 branches and each branch is ensheathed by a spathe. Each branch has 2-3 branchlets or spikes. Each branchlet is stout, cylindrical, 30-40 cm long and 2.5-4.0 cm wide. The width of the spike decreases gradually from the base to apex. The spikes are imbricated by numerous bracts. The number of spikelets in each spike ranges from 800-1000 and each spikelet has 15-20 sessile, little florets. Each flower in spikelets is subtended by small, semi-transparent cuneate bracteoles. The flowers in the spikelets are arranged in two vertically opposite rows, serrated into each other, each spikelet forming an arch. The spikelets are arranged in parallel and nearly straight rows running clockwise or anti-clockwise, around the spike. In total, a single spadix may contain 2,00,000 to 2,50,000 florets. Flowers are sessile, 3 sepals, imbricate, cuneate with truncate tips. Three petals, short, ovate and imbricate. Six stamens, filaments connate into a stalk with corolla. Anthers are large, sessile, oblong, bilocular and split longitudinally. It takes 16-25 days for the flowers to open after opening of spathes. Flowers open from the lower half of the spike and extended to both ends. Rarely more than one flower opens at a time. Pollen shedding is simultaneous with flower opening and pollen grains are ellipsoidal.

Female Spadix: The female spadix has only 2-4 branches or spikes, sheathed by spathes. The upper half of the spike is imbricated by bracts while the lower end is smooth peduncle. A barren bract ensheaths the spike, from where the flowers arise and the terminal of the spike extending to 5-8 cm beyond the flower is also ensheathed by barren bracts. The number of female flowers in a spadix ranges from 30-75. The female flowers are large, globose, perianth is 6 lobed, fleshy, imbricated, reniform and accrescent. Six staminodes, globose ovary, 3-4 celled, pistils 3-4, syncarpous, 3 stigma, sessile and recurved. Ovules are basal and erect (Sankaralingam *et al.*, 1999).

Pollination: Pollination occurs through insects (bee, wasp, beetles) and wind. It takes 120- 130 days for the fertilized female flowers to mature into ripe fruits (Sankaralingam *et al.*, 1999).

Fruit: The fruit is a fleshy drupe and weighs around 1 to 3 kg. Fruit is almost spherical with flat bottom. At the stalk end, six reniform perianth lobes are arranged in two whorls of three each. The pericarp is thin, fibrous and brittle. It is creamy when young and turns black or pinkish yellow on maturity. Mesocarp is edible, fibrous and yellowish in colour which surrounds the one to three hard coated seeds. Though the fruit develops from three fused carpels, the number of pyrenes within the mesocarp varies from 1 to 3. The endocarp is hard and it covers each seed, which has a brown testa. The endosperm is gelatinous when the seed is young, filling the entire cavity after 60-70 days of fertilization. As the fruit matures, the endosperm hardens, forming a cavity at the centre. The embryo is positioned below the germpore and embedded within the endosperm. The germpore is situated at the stigmatic end of the fruit (Sankaralingam *et al.*, 1999). Endosperm contains fats, protein, oils but no starch. 200-300 fruits are produced annually.

Area and distribution

The palmyrah palm is distributed in tropical zone of Asia, Africa and Australia. The major countries growing palmyra palm are India, Sri Lanka, Bangladesh, Myanmar, Thailand, Vietnam, Malaysia and Indonesia. This palm is found abundantly in dry or sandy localities of Kerala, Karnataka, Andhra Pradesh, Tamil Nadu, Madhya Pradesh, Chhattisgarh, Odisha and Rajasthan. According to Veilmuthu (2020), India has the highest number of palmyra (85.9 million) in the world and out of this 60% are in Tamil Nadu (51 million) and the district of Thoothukudi ranking first.

5.2 Soil and Climate

Palmyra can be grown in any type of soil; however, growth is luxuriant in deep loamy and sandy loam soils. Palm helps in reducing soil erosion and prevents shifting of sand dunes. Palm grows in fallow land and all soil types. Palm also grows well in drastic soil types like sand soil, red soil, black soil and alluvial soil. In areas with adequate soil fertility it gives extreme yield. The palm grows in the acidity of the soil pH 5.5 in Kampuciya. East Coast palm grows in the sand ridges and prevents soil erosion.

Palmyrah is a tropical crop and is abundantly grown in dry areas where rainfall is moderate to low. This palm grows well in arid regions and requires annual rainfall of less than 750 mm. It also grows well in wet conditions and does not suffer from flooding. It grows from sea level up to an altitude of 800 m. The average optimum temperature is 30° C, and can tolerate high temperature of 45° C and low temperature up to 0° C.

5.3 Species and Cultivars

Apart from *Borassus flabellifer*, there are many other species that exist under this genus *Borassus,* i.e., *B. aehiopium*, *B. dichotomus*, *B. madagascaiensis*, *B. secundiflorus*, *B. marseille*, *B. heineana*, *B. sambiranensis, B. sundaica.*

In India, there is no recognized variety. But palmyrah palms growing in Sri Lanka can be broadly classified into 2 varieties based on pigmentation of fruit skin. They are black and red skinned fruits. Black-skinned fruits have comparatively less red pigment on their skin. Red skinned fruits have variable amounts of black pigments along with very liberal distribution of red in their skin. Fruits and nut number per tree are significantly greater in this variety. But pulp weight per nut is less; sugar, starch and protein constitute 77%, 10% and 2.5% of the pulp respectively. The alkaloids, amino acids and minerals are in greater amount in red skinned varieties. The other favourable fruit features, along with the sap-yielding characteristics of these varieties, seem to favour selection of red-skinned fruit variety for commercial exploitation (Sankaralingam *et al.*, 1999).

Released variety: Palmyrah Research Station, Srivaliputhur (TNAU) has released one improved cultivar, namely, SVPR-1 Palmyrah palm. It is a semi-dwarf type with a high sap yield of 298 litres per palm in a tapping duration of 95 days. The sap of this cultivar has high jaggery content (144 g/lt of sap) and high brix content.

5.4 Propagation

Palmyra palm is propagated by seed. Selection of good mother palm is very important for getting better seed nut. Mother palm should be dwarf, high yielding, early and regular bearer, free from disease and pest. Seed nut should be collected from well matured fruit bunches indicated by the yellow tinge in the stylar end of the fruit and heaped under shade for four weeks. Separated seed nuts are loosen from fruits and soaked in carbendazim @ 0.1 % for 24 hours to reduce the tuber rot incidence and enhance the germination. The shrunken and weightless nuts should be discarded.

Raising of planting materials: Seeds may be sown *in situ*. Three to four good whole nuts are planted in the pit and covered with half-filled sand or sand soil mixture and covered with mulch. Seeds are generally sown during rainy season and it takes around 15-20 days to germinate and 150-160 days for the first leaf of the seedling to come out of the soil.

Seedlings may be raised for transplanting by sowing seeds in the sand bed nursery, keeping 10 cm spacing followed by covering with 5 cm sand. When seedlings become one year old, they should be lifted and placed in polybags. They are kept in shade for root establishment; after that they are transplanted in the main field. Seedlings may be soaked in 50 ppm NAA for better rooting. Sankaralingam (1999) reported that the treatment of seed nuts with 0.1 % carbendazim followed by 10 % cow dung slurry and 1 % Trichoderma resulted in 82% germination.

5.5 Planting Systems and Method

Generally square system of planting is preferred. Pits of 30 cm × 30 cm × 30 cm sizes are dug at a spacing of 3 m × 3 m accommodating 1110 plants/ha (450/acre). If triangular system of planting is adopted it accommodates 500 trees per ha. The pit is half filled with a mixture of FYM (10 kg) and top soil. The nut is placed with its germpore facing down ward or side wise at a 5 cm depth and 100 g malathion is sprinkled around it to prevent termite attack and nut is covered with soil. Planting is normally done during monsoon season. If there is short fall of rains, pot watering immediately after planting and at alternate days up to a month is required. Later on, watering can be done once in a week during dry periods for a year.

5.6 Gap Filling

After initial establishment of the plantation, gaps are to be filled with healthy saplings whenever necessary.

5.7 Interculture

The field is to be ploughed occasionally. Leguminous crops are preferred as intercropping as it enhances the soil fertility status. Groundnut, ginger, cowpea and greengram can be raised as intercrops during rainy season. Fruit trees like custard apple, ber, pomegranate, aonla can be planted as mixed crops.

5.8 Shade Regulation

In *theri* lands, young seedlings require shade to protect them against sun and desiccating wind, using one or two dried palm leaves.

5.9 Water Management

Irrigation is to be provided in the basin of individual palm. In drier region for efficient water management, rain water harvesting is very essential along with mulching in the basin area to conserve the moisture. To increase the sap yield and fruit yield, pitcher irrigation is recommended twice a month during tapping season (Sankaralingam and Khan, 2001).

5.10 Pruning and Handling

Defoliation and tending should be practised at the initial stages of growth of the palm. Removal of persisting leaf base periodically is known as tending. Tree cleaning is necessary once a year before tapping. Young leaves should not be removed from juvenile palms. One or two leaves can be removed when the palm reach a height of 2 m. An adult tree can be defoliated up to 50 % keeping 16-22 leaves at the crown. Removing all the leaves leaving the bud and crown is totally harmful to the palm. Foliage and butts (leaf base) should not be removed during summer. Butts and old sensed leaves are to be removed.

5.11 Nutrition

For manuring, farmers generally adopt sheep penning. However, application of FYM @ 10 kg per pit before planting nuts/seedling is recommended. The rate of application of FYM may be increased biennially up to 60 kg/pit/year. Velu (1992) reported highest sap yield (301.66 lt./palm) with application of 60 kg FYM. This treatment also resulted in longest tapping period (136 days) and highest numbers of palms available for tapping (8 out of 10). Arumugam *et al.* (1994) observed highest sugary sap yield (113.82 lt./palm) and TSS (11.02 °B) with application of 50 kg FYM alone. Application of 100 kg green manure or 500 g each of N, P and K also gave good result.

5.12 Crop Protection

5.12.1 Insect-Pests

1. Rhinoceros beetle (*Oryctes rhinoceros*)

It is one of the most serious pests of palmyra. Adult beetle bores hole and feed on the unopened frond and spathes. On opening the damaged leaves shows the geometrical cuts on the leaflets.

Control: The crown and the spathe between leaf axils are to be filled with a mixture of 250 g of malathion 4 % and sand in equal proportion. The crown is examined at regular interval and the adult beetles are hooked out and killed. In young palms, three naphthalene balls are kept at the leaf axils once in 45 days. Since the grub proliferate in manure pits, malathion 4 % dust can be applied and mixed with manure. Inoculation of baculovirus to adult beetle will protect the leaf and crown damage. Soaking castor cake in water and keeping it in pots in and around the garden will attract the adult beetle. Longitudinally split palm petioles are soaked in toddy and kept in garden to bait the beetle.

2. Red palm weevil (*Rhyncophorus ferrugineus*)

The grubs feed on the soft pith within the stem and crown, leading to drying. Patches of viscous fluid may ooze out of the bore holes on the stem. The weevil infestation is usually more during tapping months.

Control: Care should be taken not to damage the stem while removing butts or leaves. Removal and burning of wilted and damaged palm are necessary, since it may serve as the breeding ground for the weevils. Treating wounds if any, on the stem can be done with a mixture of malathion 50 % WP and tar. Root feeding with monocrotophos 10 ml in 30 ml of water during off season is required. Root feeding should be stopped 45 days prior to tapping and should not be done during tapping period April to July to prevent any harmful effect on users of neera.

3. Back headed caterpillar/ leaf eating caterpillar (*Opisina arenosella*)

Young palms are more susceptible to infestation. Larvae scrap chlorophyll and feed on them leaving only the rachis. They form the galleries on lower side of the leaves.

Control: Removal and destruction of severely damaged leaves has to be done. Periodical release of larva (Braconids and Bethylids) and pupa (Eulophids) parasites and predators are required. The larval parasites Braconids and Bethylids are released @5/tree, while the pupal parasite Eulophid is released @10/tree. Depending upon the level of infestation, these parasites are to be released two to three times at 20 days interval. On young palms, foliar application of malathion or monocrotophos @ 0.1% can be done.

4. Eriophyid mite (*Aceria guerreronis*)

Stray occurrence of eriophyid mite was recorded on the fruit bunches of palmyra tree. Externally, the mite infested nuts did not exhibit any visible damage symptoms. Yellow and brownish discolouration was observed inside the perianth. There was no observed infestation in ripened fruits.

5.12.2 Diseases

1. Leaf spot (*Stigmina palmivora*)

Young seedlings and juvenile palms are more susceptible to this disease. Small reddish brown necrotic spots of less than 0.5 cm in size appear all over the leaves. Several spots coalesce resulting in brown patches leading to drying of foliage.

2. Leaf blight

Palms at all the ages are susceptible. Spots are relatively bigger having 1.0 to 1.5 cm length and 0.5 cm in width. Spots generally appear on older leaves. They are linear in shape with grey margins and brown centres. Many spots coalesce and lead to blight.

Control: Both leaf spot and leaf blight can be managed by spraying 0.2 % copper oxychloride.

3. Bud rot *(Phytopthora palmivora)*

Spear leaves become yellow and turn brown later. Infected spear bends at its base and hangs. Due to rotting of infected tissues, the spear comes off easily when pulled. Palms emit offensive odour. If unchecked, rotting will proceed to meristem leading to wilting. The disease is noticed during monsoon season and it is favoured by the damage caused by Rhinoceros beetle.

Control: Removal of infected tissues and cleaning up the crown has to be carried out. Drenching the crown with 1% Bordeaux mixture or 0.2% copper oxychloride can be adopted. Palms can be saved if only the infection is detected at an early stage.

5.13 Harvesting and Yield

Tapping and collection of sap are the major purpose of palmyra cultivation. The palmyra palm usually starts flowering at the age of around 15 years. However, under proper care it may flower at 10-12 years. Palm yields a crop of 50-200 fruits annually in 6-12 bunches per tree. When a tree attains the height of 12-18 m, it comes to flowering (13-15 years) for padaneer (sweet sap) purpose. Average of 100-200 litres Neera is obtained over a period of 4 months from February to May. Padaneer and fruit yield are highly variable in individual palms.

Tapping: Extraction of sap and Neera/Padaneer from inflorescence is called tapping. According to sex of the palm and age of the inflorescence, different kinds of tappings are available. The palm is prepared for tapping by cutting off all leaves except 3-4 at the top and young bud; the outer covering of the part of the tree from which the leaves, and racemes emerge is also removed. Kinds of tapping vary with the sex of the palm and age of the inflorescence. In the male palm the sheath covering the young inflorescence (2 weeks old) is removed and allowed to dry for three days. The pot is tied to the inflorescence and fresh cut is given in every time. The collection of sap is done for one to one & half months. In another method, one month old inflorescence is tapped. Each male spike bearing sessile flower is pre-treated by pressing and striking the inflorescence to stimulate the flow of sap. Three to six spikes are brought together wrapped with palmyra leaves and fitted to a pot. In female inflorescence the tapper softens the tissues by hitting the young inflorescence axis with an iron rod and a fork is used to press the regions from which fruit develop. Another method of tapping is done on 2 to 3 months old inflorescence where the fruits are sliced as the tapping progress. Collection of sap is done every morning and evening or in the morning only. After the sap is collected in the morning, inside of the pot is coated with lime before using it again. A new surface is made by cutting a thin slice after each collection. Application of 20 MM aqueous solution of EDTA in the morning and evening at the cut surface of spadix increases the yield 2 to 3 times. The trees are reported to die if the operation is repeated on the same trees for the three successive years without allowing any natural bursting of buds. Usually, rest is given once in three years. Seven to eight spadices yield sap at a time in each palm. The average yield from a single tree in India is 150 litres/year. The yield of sap is higher during summer than in winter (Parthasarathy *et al.*, 2006).

The sap is mildly acidic. It contains sucrose and starts fermenting very soon. Fermentation may be delayed if lacquered pots are used for collection. Cleaning and subsequent smoking of pots and use of lime solution (1.25 g/ lt.) also delay early fermentation. Use of preservatives (sulphanilamide, silica gel charged with sulphur dioxide, paludrine, mixture of citric acid (0.2%) and sodium benzoate (0.02%) and chilling) showed good results (Anon., 1988).

Yield: (a)150 litres of padaneer/tree/year (b) 24 kg jaggery/tree/year (c) Jaggery recovery/litre of padaneer: 180-250 g of jaggery.

5.14 Post-Harvest Handling and Processing and Value-Added Products

Neera: Sap is extracted by tapping of the inflorescence. This sap is known as Neera/Toddy/ Padaneer. Neera is the most important economic produce from palmyrah palm. Neera is a very sweet and delicious sap of the palm obtained by slicing the spathe of the palmyrah (or similar palm) or scraping the tender most part just below the crown.

Neera can be consumed as liquid form or fermented to a country liquor toddy or may be processed to sherbet, sugar or jaggery. As Neera is free of sucrose, it does not cause diabetes. Previously Neera was a poor man's beverage but now its potential is known to man and hence, its demand is increasing at a serious extent. Neera is high in vitamin content and is rich in minerals as well. Neera has laxative and diuretic properties as well.

Both male and female inflorescences are tapped and female palms are reported to yield 33-50 percent more sap than the males. The flowering shoots of male inflorescences are tapped while the fruiting branches of the female trees are tapped when the drupes are very small. The male palms are usually tapped from December to February and female palms are tapped from February to March but it varies from December to June in different localities.

***Toddy*:** Toddy is actually a fermented product. Sap is collected by slicing of the tip of the unopened flower. The sap oozes out and gets collected in a small pot (earthen or metal) tied underneath. Usually, the pot is pre-inoculated with small amount of toddy for fermentation process. In 6-8 hours, toddy is formed as a result of fermentation of the sugary sap with yeasts and bacteria. Unfortunately, toddy tapers are the poorest section of the society, but it is of superior nutritional and medicinal qualities. Alcohol content in toddy is estimated to be in the range from 4 to 6%. Sucrose, glucose and fructose are major sugars present in toddy. As fermentation process takes place, the sugar is converted to ethyl alcohol during fermentation.

Tender fruit: Nungu, the soft jelly like endosperm obtained from the tender fruit of palmyrah is highly perishable and seasonal. However, nungu can be processed and preserved as value added products.

Jaggery: Jaggery is a traditional product prepared from concentration of palmyra sap and sugars. This is also known as palm *gur*. The jaggery must be prepared from the unfermented tree sap. This product has nutraceutical properties and nutritional benefits as well. It has a characteristic pleasant taste and is similar to chocolate. Jaggery is prepared from Neera and hence has similar benefits as Neera except some volatile loss during cooking. This product contains an elevated content of potassium (K^+) and hence good for diabetic and hypertension patients.

Treacle: Treacle is actually the highly concentrated sap of palmyra palm which is being prepared by heating the original sap to about 1/6th of its original volume. Thus, a thick dark syrup is formed which is viscous. The TSS is actually 65% brix and pH 7.5. The temperature should not rise beyond 107°C. Temperature above 107°C would cause caramelization and darkening of colour of treacle. Treacle can prove out to be an ideal substitute for sugar and can be used as a sweetener for sweet dishes.

***Flour from young shoot*:** Young palmyrah seedlings were collected which must be tender in nature. The tender parts are skinned, washed and cut into small slices. The slices are dried at temperature 80°C until dry. These pieces are then crushed properly to obtain flour. This flour is kept in a sealed air tight plastic jar. The flour thus obtained is having 9.52% protein, 73.42% starch, 8.16% fibre, potassium (37.67%), iron (0.13%), magnesium (2.6%), calcium (1.07%), copper (0.02%), zinc (0.03%), selenium (0.2%) and phosphorus (0.34%).

Palmyrah vinegar: Vinegar from palmyrah is prepared by acetic acid fermentation of palmyrah sap. Palmyrah vinegar is very good source of carotene, vitamins and minerals besides carbohydrates and proteins. This product can be used to cure the bladder infection, urinary tract infection, skin diseases, headaches and body pains, further it is reported to improve memory also in human beings.

Nungu: It is the tender palmyrah fruit pulp which is crystalline white in colour. It has a cooling effect.

Palm cola: It is an aerated soft drink which is prepared with 11% of palm sugar. The other ingredients being cola concentrate, citric acid and food colour. The palm sugar is added to the milk and is heated to 110-115°C to remove impurities. Cola essence should be added to the mixture after cooling at the rate of 250 ml per 1000 bottles of palm cola.

Palm toffee: Palm toffee is another important and prospective product. It is prepared by mixing palmyrah fruit pulp with sugar, skim milk powder, glucose, maida and starch as ingredients. The mixture is cooked continuously with constant stirring for 40 minutes or so. The end point is determined following drop test in water. A complex mixture is formed which is then spread in an aluminium tray. The product is smeared with butter and is cut to pieces the next day. Toffees of desirable size and shape is cut and wrapped in a butter paper.

Sugar candy: It is the crystalized product made out of palmyrah neera. Palmyrah sugar candy gives cooling effect. This product is recommended for people who are suffering from pox disease to cool their system.

***Palm sugar*:** The neera collected must be clean and free of debris. Neera is boiled in an alloy vessel adding small quantity of superphosphate. After uniform boiling the liquid is allowed to cool. Later sediments and lather are removed and heated to 110°C for 2 hours until it reaches consistency just like honey. The neera fluid is then allowed to cool and is poured into a crystallizer. Then it is centrifuged to collect sugar. Then proper packing is done.

Palmyrah tubers: Palmyrah palm does not produce tubers but tubers are actually the fruit after pulp extraction being buried in the soil. The hard lump after extraction of pulp is dumped in soil and watered vigorously. After

2-4 months, the lump develops into a tuberous structure which is rich in carbohydrate. The tubers are juicy and rich in minerals.

Palmyrah fruit: Palmyrah fruit pulp could be commercially utilized to produce food items and animal feed. The whole fruit contains about 40% of undiluted pulp which is dark yellow in colour with its characteristics flavour and bitterness. Palmyrah pulp is mixed with other fruits for making jam, cordial, cream etc. Since its pulp is bitter in taste, it is better to prepare mixed fruit jam rather than palmyrah jam separately.

5.15 Economics

Economics of single palm per year

Raw material	Quantity	Finished product	Approximate value (in Rs.)
Neera/Juice	150 lt	Gur (20 kg)	4000
Karuppatti/Jaggery/Gur	20-25 kg	-	-
Panankarkandu/Palm candy	16 kg	-	-
Matured leaves	10 kg/8 nos.	Mat 6 nos.	100
Thumbu/Coir	11 kg		-
Eeark/Ekil (Leaf ribs)	2.25 kg	Brunch 12 no	70
Naar/Fibre	16-20 kg	Basket 1 no	70
Viragu/Fire wood	10 kg	-	-

Source: Veilmuthu (2020)

References

Anonymous 1988. *The wealth of India – Raw materials*. Publications and Information Directorate. Council of Scientific and Industrial research, New Delhi- 110012.

Arumugam T, Suthanthirapandian IR and Doraipandian A1994. *South Indian Hort.*, **42**:236-238.

Mani A and Dutta P 2018. An array of processed and value-added products from palmyra palm. *Journal in Science, Agriculture & Engineering*, **8**: 154-156.

Parthasarathy VA, Chattopadhyay PK and Bose TK 2006. *Plantation Crops*. Palmyrah Palm. Vol-2, pp. 493-511.

Sankaralingam A 1999. *J. Mycol. and Plant Pathology*, **29**:114-115.

Sankaralingam A and Khan HH 2001. *Indian Fmg.*, pp15-17.

Veilmuthu P 2020. Palmyra-nature's perennial gift in the face of climate crisis; 2020. http://climatesouthasia.org/palmyra-natures-perennial-gift-in-the-face-ofclimate-crisis/

Velu G 1989. *Madras Agric. J.,* **76**: 592-598.

OUTCOME ASSESSMENTS

PART A

Answer the following questions true or false.

1. Palmyrah palm is monoecious. True/false
2. Neera can be consumed as liquid form. True/false
3. Palmyrah is grown extensively in Kerala. True/false
4. Toddy is actually a fermented product. True/false
5. Extraction of sap and Neera/Padaneer from inflorescence is called tapping. True/false

PART B

Answer the following questions.

1. The palmyrah palm is distributed in tropical/subtropical zone of Asia, Africa and Australia.
2. What is toddy?
3. What is neera?
4. Removal of persisting leaf base periodically is known as —.
5. How vinegar is prepared from palmyrah?

PART C

Write a brief note on each of the following.

1. Value-added products in palmyrah.
2. Propagation in palmyrah.
3. Harvesting in palmyrah.
4. Diseases in palmyrah.
5. Species and cultivars in palmyrah.

6

Production Technology of Cocoa

Botanical Name: *Theobroma cacao* L.
Family: Sterculiaceae
Chromosome No: 20
Place of Origin: Tropical South America
Common Name: Cacao, cacao bean, cocoa, cocoa bean

6.1 Introduction

Cocoa (*Theobroma cacao* L.) is a perennial plant and is used as beverage crop. The generic name *Theobroma* means the "Food of the God". The word cacao refers to the tree, while the word cocoa refers to the drink made from its seeds. The roasted product of the dried beans is called as *Cacao nibs* that are used for the manufacture of various products. When cacao nibs are ground, the resulting product is called *chocolate liquor or mass*, which contains around 55% fat. The fat that is pressed from chocolate liquor is termed as *cacao butter.* It is mainly used for the manufacture of chocolates, in pharmaceutical preparations and soap making. After pressing out the fat the resultant product with reduced fat content is called *cacao powder.* For preparation of chocolates cacao powder is mixed with cacao butter and sugar with a definite ratio. Milk chocolates are prepared by adding milk to the above mixture. The cocoa powder contains 22-25% fat, 20.4% protein and 35% carbohydrate with caloric value of 452 cal. Rind or shell of pod is used as a cattle feed.

There are more than 20 species in the genus but the cocoa tree (*Theobroma cacao* L.) is the only one cultivated widely. All these twenty species are native to Tropical America. The best-known other member of the family is *Cola accuminata*, which produces Cola nuts and it is the main source of the stimulating principle in making of "Cococola". Cocoa is a semi deciduous tropical plant, reaching a height of 5-8 m with a dense foliage and rounded canopy. The main trunk is short, branches at a height of 1.0 to 1.5 m from the ground level. Branches arise in whorl in a horizontal fashion. The horizontal branches are called Jorquette or fan and the process of branching is Jorquetting. The upright-growing shoots are called “chupons”—basically what we would call “water sprouts” or “suckers” on other fruit trees. Cacao exhibits a flushing-type growth habit, with two to four growth flushes per year. Cacao is cauliflorous i.e., flowers and fruits are borne on the old wood of the trunk and main branches. The flowers arise in the leaf axils (cushion). The flowers are hermaphrodite, small, reddish-white colour and odourless. The inflorescence is a compressed cymose with a short, thick peduncle. The flowers are hermaphrodite, pink or whitish, regular, five sepals, five petals, ten stamens in two whorls of which five are fertile with superior ovary.

Cocoa appears to have been introduced in India as early as in 1793 and its commercial cultivation was started in early 1960 as mixed crop in the coconut and arecanut plantations. This favourable situation, coupled with large scale distribution of planting materials could bring about an enviable area coverage recording 29,000 ha under cocoa by 1980-81. During the implementation of 11th and 12th Five Year Plan programmes there has been a large-scale distribution of hybrids and High Yielding Varieties of Cocoa which has helped increase the area and production of this crop. Cocoa is hardly grown as a mono crop. The tropical diversified congenial climate available in India provides immense scope for its wide spread cultivation. Kerala was the leading State in promoting cocoa cultivation. From 1997-98 onwards the non-traditional tracts of Karnataka and other States like Andhra Pradesh and Tamil Nadu started developing cocoa cultivation.

6.2 Soil

Cocoa can be grown on a wide range of soils and the soil should be deep with rich in organic matter. It can be grown in forest soil which is rich in humus. The loams and sandy loam soils with high C: N ratio is suitable for growing of this crop. The soil should be able of retaining moisture during summer and allowing movement of air and moisture as cacao requires regular supply of moisture for proper growth and development. The trees are more sensitive to moisture stress than other tropical crops. Shallow soils should be avoided and the minimum depth of the soil should be at least 1.5m. The soils pH of 6.0 -7.5 is ideal as major nutrients and trace elements will be available at this pH. The crop can withstand flooding but it will not tolerate stagnant, water logged conditions. A minimum requirement of 3.5% organic matter say 2% Carbon in the top 15cm is ideal for growing cocoa plantation. Shallow soils should be avoided. Cocoa does not come up in coastal sandy soils where coconut flourish.

6.3 Climate

Cocoa thrives between 20°N and 20°S. It is a low altitude crop (Elevations 200 – 300 m MSL) but can be cultivated up to 700 m above mean sea level. Cocoa is a crop of humid tropics and requires well distributed rainfall with an annual precipitation of 1500 to 2000 mm. This crop can also be grown in other regions by substituting the rainfall with irrigation during dry periods and for suitable cultivation. Care should be taken in dry periods and it should not exceed 3 to 4 months. Cacao tolerates a minimum temperature of 15°C and a maximum of 40°C, but temperature around 25°C is considered as optimum. It cannot be grown commercially in areas where the minimum temperature fall below 10°C and annual average temperature is less than 21°C. Humidity is uniformly high in cocoa-growing areas, often 100% at night, falling to 70-80% by day, sometimes low during the dry season. The most marked effect is on leaf area, plants growing at low humidity (50-60%) having larger leaves and greater leaf area than plants growing at medium (70-80%) and high (90-95%) humidity; under the latter conditions leaves are small and tend to be curled and withered at the tip.

6.4 Choice of Cultivars

There are three major varietal groups, namely, Criollo, Forastero and Trinitario. Among them, Forastero is the one that is commercially grown all over the world. It is high yielding, more resistant to pest and diseases, and more tolerant to drought compared to Criollo. The comparison between Criollo and Forastero is as follows:

Character	Criollo	Forastero
Cotyledons	White when fresh and turn cinnamon coloured on fermentation	Flat and purple when fresh and turn dark chocolate brown on fermentation
Pod colour and no. of beans/ pod	Dark red, 20-30	Yellow, >30
Other pod characters	Soft fruit, rough surface with 10 distinct longitudinal furrows. Ridges prominent and irregular, pod round with pronounced point	Smooth, inconspicuous ridges, thick and hard walled, melon shaped with rounded end
Flavour and aroma	Sweet taste and bland flavoured	Harsh flavour, bitter taste
Duration of fermentation	3 days	6 days
Others	Susceptible to drought, tender so most suitable as intercrop in coconut and arecanut gardens, susceptible to diseases	Less susceptible to diseases

Character	Criollo	Forastero
Yield	Comparatively low	Comparatively high
Adaptability in India	Poor adaptability hence discouraged for commercial cultivation	Good adaptability and hence recommended for commercial cultivation

Some of the important cultivars developed are suitable for commercial cultivation in Kerala, Karnataka, Tamil Nadu and Andhra Pradesh that are furnished below:

Cultivar	No. of pods / tree	No. of wet beans / pod	Single dry bean weight (g)	Dry bean yield (kg / tree)
CCRP I (M 16.9)	56	46	0.8	2.06
CCRP 2 (M 13.12)	90	45	0.8	3.24
CCRP 3 (GI 5.9)	68.5	42	0.8	2.28
CCRP 4 (GII 19.5)	66	42	0.8	2.22
CCRP 5 (GIV18.5)	38	42	0.8	1.28
CCRP 6 (GVI 55)	50	48	1.1	4.56
CCRP 7 (GVI 56)	78	47	0.9	3.23
CCRP 8 (PII 21)	90	48	0.9	3.88
CCRP 9 (SIH 7.1)	105	36	0.8	3.02
CCRP 10 (SIIH 4. 13)	79	41	1.1	3.56
VTLC1 (I-14)	56	45	1.17	2.51
VTLC5 (II-67)	51	40	1.1	2.51
VTLC8 (III-105)	56	42	1.06	2.0
VTLC 9 (III-35)	66	42	1.09	3.0
VTLC 11 (I-56)	61	42	1.2	2.0
VTLC 30 (NC-42/14)	43	40	1.08	1.54

6.5 Propagation

Cocoa can be propagated from seeds or by vegetative means.

6.5.1 Seed Propagation

This method is commercially practiced and for this seed pods are generally collected from trees yielding 80 to 100 pods per year. The seeds are collected from pod weighing 350-400 g. Cocoa seeds lose their viability within a week of harvest from pods. So soon after the harvest from the pods, the seeds should be sown immediately. Viability of the beans can be extended for some more days if freshly extracted seeds are stored in moist charcoal and packed in polybags. The seeds are rubbed with dry sand or wood ash to remove mucilage. The beans are planted with their pointed end upwards, either in plastic bags (25 x 15 cm size, 150 gauge) or in raised beds. If sown in beds, young seedlings are usually transplanted into polythene bags after about two weeks of germination. The seedlings are ready for transplanting in the field after about 3 to 4 months or they attain a height of 30 cm. In trees of Forastero type, green colour pods are selected when they are immature, having smooth or shallow furrows on the surface without prominent constriction at the neck.

6.5.2 Vegetative Propagation

Cocoa plants raised from seedlings show great variability. So vegetative method of propagation through cuttings, soft wood grafting, forkert method of budding etc. can be adopted for propagation.

6.5.2.1 Semi-Hard Wood Cutting

Cuttings are obtained from recent flushes and they are made of size 15 cm long and three to four terminal leaves that are treated with a mixture of 4% IBA and 0.2% NAA. Then, the cuttings are planted in polythene bags containing coir dust. By adopting this method around 60 to 70% success was achieved.

6.5.2.2 Budding

The bud wood is normally collected from chupons because buds that are collected from the fans may develop into bushy type. The brown bark shoots are selected in which just hardened leaves are selected as bud wood. The fresh bud wood can be used for budding but the percentage of success is more in case of buds that are collected from pre-curing shoots. Pre-curing consists of removal of the lamina portions of all the leaves from which the bud stick is chosen. The petiole stump falls off within 10 days and the buds begins to grow. After extraction of bud wood, wash gently and dip in benzyl chloride. The cut ends are sealed using molten wax. The bud wood is generally wrapped in moist cotton wool/wet tissue paper/ blotting paper in order to avoid wilting and kept in boxes for future use. Six to twelve months root stock is selected for budding. Budding methods like T, inverted T, patch and modified Forkert methods can be adopted. In Patch budding, a patch of about 2.5 cm length and 0.5 cm width is removed from the rootstocks and a bud patch of 2.5 cm length and 0.5 cm width is collected from the desired bud wood plant and then inserted into the rootstock. A polythene tape is used for tying firmly. Above the bud a vertical cut is made and when the bud starts to grow as shoot, the upper portion of the rootstock is cut above the bud. Under normal conditions, success is 70-90 per cent.

6.6 Land Preparation and Planting

The land is prepared by ploughing and harrowing and soil is brought to a fine tilth. Pits of 75 cm^3size are prepared and filled with a mixture of soil and FYM. Planting should coincide with the onset of the monsoon but in places where irrigation facilities are available planting can be done throughout the year. It is planted at a distance of 2.5-3.0 m both between and within rows. The planting may be taken up either in the beginning of the monsoon, i.e., in May-June or at the end of the South-West monsoon, in September. Cocoa thrives best under partial shade. When this crop is grown as monocrop, provide shade by planting of shade plants like gliricidia, sesbania and banana. Dadap (*Erythnina lithosperma*) can be planted at 3 m× 3 m spacing to provide shade. Dadap needs pruning every year. For more permanent shade, *Albizzia stipulate* can be planted adopting 9 m ×9 m or 12 m × 12 m spacing. In Kerala, Karnataka and Tamil Nadu the Cocoa is grown as intercrop under arecanut and coconut garden. In arecanut garden the spacing of arecanut should not be less than 2.7 m × 2.7 m. In coconut gardens, it can be planted 2.7 m apart in a single row. Under the double hedge system, cacao is planted in two rows adopting a spacing of 2.7 m within the row and 2.5 m between rows where the coconut is planted at a normal spacing of 7.5 m × 7.5 m. Economic analysis showed that additional income from inclusion of cocoa as mixed crop was to the tune of 69%.

6.7 Nutrition

Cocoa plant requires correct amount of nutrients for its proper growth. Application of 25-30 kg cattle manure per annum is recommended for initial establishment better root growth. In addition to organic manuring, the plant needs to be fertilized with chemical fertilizers, i.e, 100 g N, 40 g P_2O_5 and 140 g of K_2O per tree per year. In the first year of planting, the plants may be given one third of the above dose, while two third of dose is to be applied in the second and third year, and fourth year onwards full dose of fertilizers is applied. While applying the manures and fertilizers, care should be taken to avoid serious damage to the surface feeding roots. The young plants are fertilized in circular trenches and adult pants can be broadcasted in the root zone (up to a radius of 120 cm**)**. The fertilizer is applied in two splits, the first dose in March-April and the second dose in September-October. Application of VAM (Vasicular-Arbuscular Mycorhiza) @ 50 ppm per plant is found to be beneficial. Foliar application at 0.3% zinc sulphate plus 0.15% lime is suggested to control the problem of zinc deficiency (Chlorosis of leaves in patches, young leaves becoming narrow and sickle shaped, twig dieback).

6.8 Mulching

In the initial establishment of the plant, it is essential to keep the plant basin free from weeds. Mulching with materials like dry paddy straw, coconut husk, cocoa husk, rice husk etc. not only helps to conserve moisture but also repress the weed growth. A mature cocoa plantation with a proper canopy growth is sufficient to prevent weed growth.

6.9 Shade Regulation

Approximately seventy per cent (70%) shade is absolutely necessary for newly established young cocoa plants as it reduces the early mortality of seedlings due to exposure to direct sunlight. This requirement is gradually reduced to twenty-five per cent (25%) shade for cocoa trees over five to seven years. Shade also reduces water loss in the dry season. In the case of active production period no shade is needed if the cocoa trees are getting sufficient nutrients and water throughout the year. Since cocoa fields are not normally irrigated in the dry season some permanent shade approximately twenty-five per cent (25%) is needed for fruit bearing cocoa trees. Two types of shade providing trees may be used in cocoa production, i.e., temporary shade and permanent shade.

A. Temporary Shade: Temporary shade trees provide shade for young cocoa plants until such time that the permanent shade trees are established. Temporary shade is usually provided by food crops which give the farmer an income until the cocoa trees mature and start production. Plant temporary shade trees at least four to six months before the young cocoa trees are planted. These will provide shade for young cocoa plants. Maintain for a period of about one to five years.

B. Permanent Shade: Establish permanent shade trees to form a canopy over the more mature cocoa plants. Plant these at least one year before the young cocoa trees are planted. As permanent shade *Albizzia stipulatus* are planted at spacing of 9 m × 9 m or 12 m × 12 m. In South India two types of shade are recommended. Papaya and banana can be taken before planting as primary shade whereas *Gliricidia* and *Indigofera* planted as secondary shading plant in South India.

Management of Shade: Adjust the shade as cocoa trees mature. During the first year remove temporary shade to allow 50% of the total light to pass through. Progressively remove the temporary shade to allow up to about 25% shade as the cocoa trees develop and their branches overlap to form an unbroken canopy. Remove all temporary shade by the fifth year. Some permanent shade (25%) is generally needed for bearing cocoa trees. Reduce shade in mature fields depending on the availability of water.

6.10 Weed and Water Management

Weeds compete with the main crop for moisture, nutrients, light and thereby affect the growth and yield of cocoa. Weeds reduce the yield potential of cocoa plant as there is decrease in photosynthetic capability and nutrient reserves in the leaves. In the early stages of growth, moisture stress is particularly critical to the young cocoa seedlings. To reduce the stress, the cocoa plants defoliate their lower leaves. It is therefore, important that weed competition should be minimal during early growth stages of the cocoa plant. In a well-established cocoa plantation from the third year onwards plant canopy coverage is significant and the heavy shade greatly reduces weed growth. Weed problem further reduces in the fifth year as thick canopy coverage increases and a layer of cocoa litter has been built up.

Cocoa plants need continuous supply of moisture for optimum growth and yield. If there is no adequate water supply during summer months, the yield will be reduced. During summer the plants have to be irrigated at weekly intervals. Similarly, under mixed-cropping systems, moisture stress affects the yield of both the crops. In mixed cropping, supply of adequate irrigation increases the yield of both corps by 30%. The young plants should be irrigated frequently depending on the moisture availability in the soil.

6.11 Pruning

Pruning in cocoa is a regular phenomenon carried out to develop a good shape. The cocoa plant generally grows in different tiers and for harvesting and other operations the height is maintained to a convenient height, i.e., 1.0 m to 1.5 m. It is advisable to restrict the growth at first tier or 2nd tier by periodical removal of vertical growth. The chupon or vertical growth of the seedlings terminates at the jorquette, where four or five fan branches develop. The second jorquette can be permitted to build up, if the first jorquette is formed very close to the soil. Generally, 3-5 fan branches are developed at each jorquette. When more fan branches develop, one or two weaker ones may be removed. The pest and disease affected branches should be removed from time to time.

6.12 Top Working

Top working is a method which helps in the rejuvenation of old and unproductive cocoa plantations by changing old poor yielding cocoa plants to high yielders. A poor yielding cocoa tree of any age can be transformed to a high yielder by the simple procedure of top working. The method is similar to the budding on seedlings. The tree to be top worked is snapped back just below the jorquette (1-1.5 m above the ground) after cutting half way through the width. Patch budding is done on three or four newly formed most vigorous chupon shoots and the rest of the chupons are removed. Budding should be done only when the shoots attain pencil thickness and their leaves are hardened. The bud wood is taken only from fan shoots of high yielding trees. Patch budding can be easily done on these shoots by removing a patch of bark of 2.5 m length and 0.5 cm width and adding a patch of bud of similar size. This scion is fixed in position and protected by firmly tying with a polythene tape. Three weeks after budding the tape is cut off and the stock portion of the seedling above the bud union is snapped back. The snapped portion is removed only after at least two hardened leaves develop from the bud. When sufficient shoots are hardened the canopy of the mother tree can be totally removed. Top working can be done during all seasons. Still, it may be more convenient if this operation is done in a rain-free period in irrigated gardens. For rainfed situations, this may preferably be done after the receipt of pre-monsoon showers. Top worked trees grow much faster than budded plants of the same age especially because of the presence of an established root system. They start yielding heavily from the second year onwards while a budded cocoa plant of the same age may take five years for the same.

6.13 Crop Protection

6.13.1 Pest Management

Like many other fruit plants cocoa is also subjected to the ravage of insect pests at various growth stages. As many as fifty species of insect pests have been recorded worldwide. However, only few of them causes economic loss to the crop. A brief account of the major insect pests attacking cocoa and their management is given in the following paragraphs.

A. Stem Borer *(Interbela tetronis* and *Interbela guadrinotata)*

Both the species of stem borers are polyphagous and distributed in India and many other countries. The caterpillar eats the bark and bore into the stem near the junctions of the branch. The main host plants are mango, citrus, litchi, ber and cocoa. The eggs are laid in crevices in bark during May-June and the newly hatched larvae tunnel into the bark and feed during night. The feeding tunnel is filled with frass and faecal pellets. Larval period extends for several months. Then they pupate and adults emerge in the following May. So, the insect completes one generation in a year.

The red stem borer of coffee (*Zeuzera coffeae* Nietn) also attack the cocoa trees. The portion of the branch above the point of the entry of the pest dries up. Control of the pest is best achieved by pruning off and destroying of attacked branches by local application of carbaryl (0.05%). Injecting kerosene/petrol in the boring holes and plugging with mud is also effective against the pest.

B. Mealy Bugs (*Planococcus lilacinus* and *P. Citri*)

Mealy bugs are emerging insect pests of cocoa in India mostly during summer season. Feeding on the tender apical shoots by adult females and young ones results in stunted growth and deformation of the affected parts, withering of seedlings which ultimately dry up. Feeding on the rind of the pods results in irregular cracks and pitting. They also cause cushion absorption and wilting of young developing pods (cherelles). The population build up of the bugs is more during summer months usually increasing from April-May till the onset of monsoon. Spraying of dimethoate @ 2ml/litre in the initial stage of the infestation and repeating at 25-30 days interval controls the pest infestation.

C. Tea Mosquito Bug (*Helopeltis* spp.)

The important species of the tea mosquito bug found in Asian region belongs to genus *Helopeltis* (*H. antonii, H. therora and H. bradyii).* These species are also distributed in Sri Lanka, Indonesia and Malaya. *H. bradyii* is commonly observed infesting cocoa. Nymphs and adults of this mired bug suck sap from the leaves, young shoots, inflorescences and pods. Due to sucking of the plant fluid, resinous gummy substance oozes out of the feeding punctures, which finally turns black and crinkling of tender leaves. Black lesions on the affected shoots cause drying of the immature pods. Deformation of the pods occurs because of multiple feeding injuries. Severe infestation shows scorched appearance leading to the death of shoots as well as growing tips. Foliar application of endosulfan (0.05%) is recommended for the control of tea mosquitoes attack.

D. Cocoa Fruit Borer (*Conogethes punctiferalis*)

This minor pest is assuming major one now-a-days. Caterpillars soon after hatching from eggs feed on rinds/husk of cocoa pods. Later they bore in and feed on the inner content of the pod. Extruding of granular faecal materials of pellets are commonly seen on the outside of the pods. Spraying of rogor @ 0.05% on the infested plant reduce the infestation.

E. Aphids (*Toxoptera aurantii* B.de F.)

Light green to brown coloured aphids colonize in the terminal shoots, succulent stems, flower buds and small cherellies. Severe infestation causes premature shedding of flowers and curling of the terminal leaves. In the seedling stage they attack the terminal 3-4 leaves. Being sap feeders, they continuously suck nutrient from the plants resulting in withering of apical shoots and stunted growth. Foliar application of 0.025% dimethoate @ 25-30 l /ha can control the pest infestation. On need basis further spray can be given at 15-20 days intervals.

6.13.2 Disease Management

A. Black Pod Rot Disease (*Phytophthora palmivora, P. citrophthora, P. capsici*)

Black pod rot of cacao is caused by a pathogen in the genus, literally translated as the "plant destroyer. There are more than 80 species of *Phytophthora* that cause plant diseases, of which *Phytophthora palmivora* is important; this is responsible for pod loss of 20 to 30%. A small brownish spot appears at the point of infection. Such infections can begin at the stem- or blossom-ends of fruits. Infection spreads rapidly across the outer surface of the pod, covering the entire pod in a few days and the infected areas turn from brown to black. The pathogen moves deeper into the pod, infecting and destroying the beans. Cankers can form under the bark of infected stems and branches. There may be a dark spot on the bark that oozes reddish fluid.

A site with relatively low rainfall and good drainage is recommended. Infected pods should be removed from the area and destroyed. A single infected pod has the potential to release 4 million sporangia. Trees should be spaced and pruned to allow for increased airflow in and around the orchard. Chemical control is an option to control the disease. Application of Bordeaux mixture before onset of southwest monsoon and monthly interval application can reduce the incidence of the disease.

B. Phytophthora Canker (*Phytophthora palmivora*)

The cankers appear either on the main trunk, jorquettes or fan branches. Water-soaked lesions are first seen on the stem portion with dark brown to black margin. A reddish-brown liquid also oozes out from these affected portions and later it dries up. Later the affected stem and branches wilt. In severe conditions, the lesions completely cover the stem hindering water and food circulation in the plant tissues, thereby death of the tree. It can be controlled in the initial stages by removal of diseased bark portion, then the wounds need to be dressed with Bordeaux mixture or copper oxychloride paste.

C. Pink Disease (*Corticium salmonicolor*)

The first sign of the disease is white threads of the fungus over the bark and these white pustules appear at cracks and natural openings present in the bark. Later, the fungus forms a pink crust which produces spores. The pink colour represents profuse conidial production by the fungus. Later in occasional cases the colour fades and orange-red pustules are seen with another type of spore. The fungus penetrates and kills the bark, cracks appear, the crust becomes grey and sunken and gum may be present. The leaves die on the infected branches, but remain attached. When it is too dry and unfavourable for growth, the fungus remains alive in the branch and trunk cankers. Apply Bordeaux paste at the fork region and at the cut ends of the twigs and spray with 1% Bordeaux mixture before the onset of monsoon. Repeat spraying again once or twice during the monsoon season according to the intensity of the disease.

D. Vascular Streak Dieback (VSD) (*Oncobasidium theobromae*)

The first symptom of the disease is yellowing of leaves. Yellow leaves with green spots and abnormal development of axillary buds, on a cocoa branch affected by vascular streak dieback, *Oncobasidium theobromae*. These disease-affected leaves fall from the tree. When the infected shoot is split lengthwise, there is always a characteristic brown streaking. All affected branches are pruned off 30 cm below the last point of visible vascular streak of the stem to prevent further spread within the plant. Judicious within canopy pruning improves aeration and sunlight penetration, and reduces the disease spread. Apply Bordeaux paste on the pruned area and spray 1 per cent Bordeaux mixture or copper oxychloride 0.25 per cent twice, i.e., in May-June and again in October as a prophylactic measure. Soil drenching with 0.25 per cent copper oxychloride (2.5g/ litre of water) around the plant. Grow VSD resistant and high yielding cultivars like CCRP-1 to CCRP-7.

E. White Thread Blight (*Marasmius scandens*)

The fungus spreads longitudinally and irregularly in young branches of the affected plants. The diseased leaves turn to dark brown and get detached from the stem. The fungus invades the cortical tissues which eventually turn dark brown. The extensive death of the young branches with suspended leaves in a row is the common field symptoms of white thread blight. The damage can be reduced by removal of the dead materials and pruning of affected parts. Shade management is necessary to control the disease. The cut end of the branch should be pasted with COC at cut ends. Apply *Trichoderma viride* and spray 1% Bordeaux mixture to control the disease.

6.14 Harvesting

Harvesting refers to removing the ripe pods from the trees and opening of the pod to extract the wet beans. Cacao flowers from the second year of planting and the development of pods is initially slow and takes about 140 to 160 days to full ripen. At maturity, the green pod turns yellow or orange colour pod turns to dark red colour with traces of orange spots. There are two main seasons to harvest the crop in a year, i.e., September-January and April-June. Under irrigated conditions pods are seen almost all through the year. The mature pods remain on the tree without damage up to a maximum period of one month. In the harvesting season pods are plucked at regular intervals of 10 to 15 days without damaging the flower cushion. The pods are harvested with the help of a knife by cutting the stalk from the plant. Then the seeds with mucilage are harvested from the pods with the help of the wooden mallets. The damaged, unripe and infested pods are separated out to ensure better quality of beans after processing. Each pod consists of 30-45 seeds surrounded with white pulp. The beans are collected and kept for a minimum period of 2-4 days for fermentation.

6.15 Processing

6.15.1 Fermentation

Fermentation is done immediately after collecting the beans from the pods. The beans are fermented to remove the mucilaginous pulp and to develop chocolate flavour, reduce bitterness, loose viability and loosen the testa, and to enable the cotyledons to expand. The factors that affect the quality of fermented beans are ripeness of pod, crop season (influence of temperature and atmospheric moisture), cultivar, fermentation method, duration of fermentation and frequency of mixing. In case of over fermentation, little chocolate flavour and acidity occur whereas due to under fermentation, increased bitterness and astringency result. The beans that are from unripe pods cannot be fermented. It was observed that beans of Criollo type ferment more quickly than those of Forastero. During the early stages of fermentation, heat is produced by the action of anaerobic microorganisms. The beans are killed by the combined effect of heat and acetic acid and the cocoa aroma and flavour potential are developed.

6.15.1.1Box Method

In this traditional method, boxes of various shapes and sizes are used. This method is adopted where the quantity of seeds is 40 kg or more. The maximum size of one is of 60 cm × 60 cm × 60 cm of 2.5 cm thick boards. In the bottom holes are made of 1 cm diameter at spacing 10 cm apart. Three number of boxes are arranged in a tier or single row so that beans can be transferred from one box to the another. In order to carry out the fermentation process the beans are placed and the top was covered with banana leaves or gunny bags. Then the beans are uncovered and transferred to the second box and third box after another 2 days interval. On the sixth day, the process of fermentation is completed and beans can be taken out for drying.

6.15.1.2 Tray Method

This method is used only for fermenting forastero cacao beans. The normal size of the wooden tray is 90 cm × 60 cm × 13 cm with a capacity to hold about 40-45 kg wet beans. This method can be adopted for fermentation of large quantity of beans and for this, large wooden boxes can be used which can hold 200-500 kg wet beans. The bottom of the tray is drilled for making 1cm holes at 4 cm apart. Six such trays are stacked one over the other for successful fermentation and one empty tray is kept to collect the drainage material. After filling all the trays, the top most tray is covered with banana leaves or sacks. The fermentation is faster here and is completed in about 4 to 5 days and the beans can be taken out for drying. This method is more convenient, time and labour saving than other method. The large growers adopt this method as the trays can be easily handled and no mixing of beans is required during fermentation period.

6.15.1.3 Basket Method

In this type small sized woven bamboo or cane baskets are used to carry out the fermentation process and small quantities (up to 10 kg) of wet beans are needs to be fermented. Bamboo or cane basket of suitable size is taken and one or two layers of banana leaves are placed at bottom and the freshly harvested beans are kept. Then the upper surface is covered with banana leaves. After one day, the basket is covered with thick gunny sacks. The beans are mixed properly and again covered with banana leaves for fermentation and the process will be completed on the sixth day and then beans are kept for drying.

6.15.2 Judging the End Point of Fermentation

The well-fermented beans need to be plumy in nature, filled with a reddish-brown exudate and the testa must remain loosened from the cotyledons. When cut opened, the cotyledons will have a bleached appearance in the centre with a brownish ring in the periphery. When 50% of beans in a lot shows the above signs, it can be considered as properly fermented.

6.15.3 Drying

On completion of fermentation, beans need to be dried in order to maintain 6 to 8% moisture. The beans can be dried in the sun or by artificial means. Sun drying can be done by spreading the beans in thin layers and stirring the beans from time to time. Under sunny days the drying can be completed in five to six days and sun-dried beans show better quality than artificial method-dried beans. In artificial method, mechanical driers are used for drying the beans and it is done by circulation of hot air around the beans in vacuum. By using the moisture meter, the moisture content of the bean is measured. The bean dryness can be determined by taking a sample in the palm of the hand and listening to its characteristic sound of dried beans.

6.15.4 Storage

The dried beans are packed in polythene bags or polythene lined gunny bags of 200-300-gauge thickness in order to maintain the quality of the cured beans. Beans should be cleaned off properly and flat, broken and other defective beans are kept separately before storing. The store house must be sufficiently ventilated and the humidity should not exceed 80% in order to prevent mould development and pest incidence in the beans. After proper cleaning and disinfection of the store house, the bags must be kept on a wooden platform with air space of about 15-20 cm. Special care should be taken not to keep other food stuff, pesticides smoke etc. as cocoa beans can absorb and retain the odour permanently.

References

Adeyemi AA 2000. Evaluation of the formulated mixture of glyphosate and terbuthylazine in the control of weeds in mature cocoa plantation. *Niger J Tree Crop Res,* **4** (2): 62–68.

Alvim P and de T 1966. Factors affecting the flowering of cocoa tree. *Cocoa Growers Bulletin,* **7**: 15 – 19.

Arunkumar K, Jegadeeswari V, Balakrishnan S, Jeyakumar P 2019. Evaluation of plus trees in cocoa (*Theobroma cacao* L.) for growth, flower, yield and yield contributing characters during the initial growth phase. *Electronic Journal of Plant Breeding,* **10** (1): 318 -323.

Balasimha D 2009. Effect of spacing and pruning regimes on photosynthetic characteristics and yield of cocoa in mixed cropping with arecanut. *J Plantation Crops,* **37**(1): 9–14.

Baligar VC and Fageria N 2005. Soil aluminum effects on growth and nutrition of cacao. *Soil Science & Plant Nutrition,* **51**: 709–713.

Beer J, Muschler R, Kass D and Somarriba E 1998. Shade management in coffee and cacao plantations. *Agro for Syst,* **38**:139–164.

Chowdappa P, Biddappa and Sujatha S 1999. Efficient recycling of organic wastes in arecanut (*Areca catechu* L.) and cocoa (*Theobroma cacao* L.) plantations through vermicomposting. *Indian J Agric Sci,* **69**: 563–566.

Elain Apshara S, Bhat VR, Ananda KS, Nair RV and Suma D 2009. Evaluation and identification of high yielding trees in Nigerian cocoa germplasm. *Journal of Plantation Crops,* **37**(2): 111-116.

Elain Apshara S and Nair RV 2011. Genetic analysis in Cocoa (*Theobroma cacao* L.) collections obtained from Nigeria. *Journal of Plantation Crops,* **39** (1): 153-156.

Jaganathan D, Thamban C, Jose CT, Jayasekhar S, Chandran KP and Muralidharan K 2015. Analysis of organic farming practices in cocoa in India. *Journal of Plantation Crops,* **43** (2): 131 -138.

Minimol JS, Suma B, Dayalakshmi EM and Jayasree PA 2014. Analysis of genetic parameters of selected cocoa (*Theobroma cacao* L.) hybrids. *International Journal of Tropical Agriculture,* **32**: 825-828.

Motamayor JC, Lachenaud P, Mota SJ, Monteiro WR, Lopes UV and Clement D 2009. Genetic improvement in cocoa. In: Breeding Plantation Tree Crops: Tropical Species (Eds.) S.M. Jain and P.M. Priyadarshan, Springer, New York., 589-626.

Nair RV, Appaiah GN and Nampoothiri KUK 1990.Variability in some exotic accessions of cocoa in India. *Indian Cocoa, Arecanut and Spices Journal,* **14** (2): 49-51.

Neela B, Frances B, Elzbieta S and Marek S 2015. Cocoa agronomy, quality, nutritional, and health aspects. *Critical Reviews in Food Science and Nutrition*, **55**(5): 620 – 659.

Olalya AO, Atayese MO and Lawal IO 2012. Effects of spacing and intercropping on the rate of infestation of parasitic weed on cocoa plantation. *Niger J Hortic Sci,* **17**: 187–182.

Prasannakumari AS, Nair RV, Lalithabai EK, Malika VK, Minimol JS and Koshy A 2009. Cocoa in India. Kerala Agricultural University, India., 72.

Prabha KP and Chandramohanan R 2011. Occurrence and distribution of cocoa (*Theobroma cocoa* l.) diseases. *Indian J. Res*. ANGRAU, **39** (4): 44-50.

Raja Harun RM and Hardwick K 1988. The effect of different temperatures and water vapour pressure deficits on photosynthesis and transpiration of cocoa leaves. In: Proceedings of the 10th International Cocoa Research Conference 1987, pp 211–214.

Ravi B and Sujatha S 2011. Nutrient uptake pattern in cocoa. Proceedings of Seminar on Strategies for enhancing productivity of cocoa CPCRI, RS, Vittal, In, pp 86–88.

Silva NPJ-da, Rocha NOG-da and Kato OR (2007) Cocoa tree growth and production in agroforestry systems according to weed management. *Rev Cienc Agrar,* **48**: 99–112.

Sujatha S and Ravi Bhat 2013. Impact of drip fertigation on arecanut–cocoa system in humid tropics of India. *Agroforest Syst,* **87**: 643–656.

Tchouatcheu GAN, Noah AM, Lieberei R and Niemenak N 2019. Effect of cacao bean quality grade on cacao quality evaluation by cut test and correlations with free amino acids and polyphenols profiles. *J Food Sci Technol,* **56** (5): 2621–2627.

Velappan E 1995. Cocoa production and development programmes. *Indian Cocoa, Arecanut and Spices Journal,* **19**:131-132.

Vikraman Nair R, Mallika VK, Amma Prasannakumari S, Koshy A and Balasubramanian PP 2000. Cocoa Cultivation: Science and Technology. Directorate of Cashewnut and Cocoa Development (DCCD), Cochin., 82.

Wessel M 1985. Shade and nutrition of cocoa. In: *Cocoa*, 4th ed., 166–194 (Eds Wood GA R and Lass RA). Essex: Longman Scientific and Technical.

Woomer PL, Martin A, Albrecht A, Resck DVS and Scharpenseel H 1994. The importance and management of soil organic matter in the tropics. In: *The Biological Management of Tropical Soil Fertility*, 47–80 (Eds Woomer PL and Swift MJ). Chichester, UK: Wiley-Sayce Publications.

Young AM 1984. Flowering and fruit-setting patterns of cocoa trees (*Theobroma cacao* L.) (Sterculiaceae) at three localities in Costa Rica. *Turrialba,* **34**:129–142.

OUTCOME ASSESSMENTS

PART A

Answer the following questions true or false.

1. The generic name *Theobroma* means the "Food of the God". True/False
2. The cocoa powder contains 10.4% protein. True/False
3. Rind or shell of pod of cocoa is used as a cattle feed. True/False
4. The branches are called Jourquetts or fans. True/False
5. In tray method, 8-10 trays are needed for successful fermentation. True/False

PART B

Answer the following questions.

1. There are more than — species in the genus *Theobroma.*
2. What is chupon?
3. Why fermentation is required in cocoa?
4. There are three/four major varietal groups in cacao.
5. Pruning in cocoa is a regular/casual phenomenon carried out to develop a good shape.

PART C

Write a brief note on each of the following.

1. Fermentation of beans in cacao.
2. Propagation in cocoa.
3. Top working in cocoa.
4. Compare forastero and criollo group of cultivars.
5. Harvesting in cocoa.

7

Production Technology of Cashewnut

Botanical Name: *Anacardium occidentale* L.
Family: Anacardiaceae
Chromosome No: 42
Place of Origin: Brazil
Common Name: Cashew, cashew tree, caju

7.1 Introduction

Cashewnut (*Anacardium occidentale* L.) is an important tropical tree crop which leads the edible nuts in international trade with 29% market share only after hazelnuts (35%) and followed by walnuts (21%) and almonds (16%). It is one of the major exports earning crops which accounts nearly Rs. 2515 crores contributing about 1.5% of the total Indian exports. It was only during the past two decades it gained the status of horticulture crop. India is the world's largest producer of cashewnut. The cashew cultivation area in India during 2018-19 was 10.89 lakh hectares with an annual production of 743 thousand metric tons of fruits. Maharashtra is a much-advanced state for horticulture which occupies large area (1.9 lakh hectares) under this crop (Directorate of Cashew & Cocoa Development, Cochin, Kerala). Cashew exports amounted to over 566 million U.S. dollars from India in fiscal year 2020. This included cashew kernels and cashew nut shell liquid. India's main export markets for this commodity included South Korea, China, Spain, United States and Belgium (Department of Agriculture, Cooperation & Farmers Welfare).

7.2 Origin and Distribution

The cashew is believed to be a native of Brazil. It was one of the first fruit trees from the New World to be widely distributed throughout the tropics by the early Portuguese and Spanish adventurers. It was introduced into India from Brazil by the Portuguese in the 16th century and it probably reached the East African coast and Malaya about the same time. It has become naturalized in many tropical countries, particularly in coastal areas. The different names of cashew in Indian languages are derived from the Portuguese name Caju which in turn originated from Acaju, the name given to cashew by the Tapi

Indians of Brazil. Other major cashewnut producing countries are Vietnam, Mozambique, Tanzania, Brazil, Kenya and Madagaskar. In India the chief producing areas are Andhra Pradesh, Goa, Karnataka, Kerala, Maharashtra, Odisha, Tamil Nadu, West Bengal, Tripura and Pondicherry.

7.3 Botany

It is an evergreen, low spreading tree and it grows up to 10-15 m. It has strong tap root and extensive lateral roots (Rhamniferous root system). It has intensive and extensive types of branching. Intensive shoot grows to 25 to 30 cm and terminates into a panicle and 3-8 laterals arise from below the panicle. It leads to bushy growth. Extensive shoot grows to 20-30 cm; bud sprouts below and leads to further growth; continues for 2 to 3 years without flowering. It is a spreading tree. High yielders have more than 60% intensive branches whereas low yielders have less than 20% intensive branches. It is a polygamous monoecious tree; 95% of them are staminate flowers and the rest are hermaphrodite. Flowering occurs in three phases, i.e., male, mixed and female phases. Pollination is carried by insects and wind; 85% of perfect flowers are fertilized of which only 4 to 6% reaches maturity. Cashew apple is fleshy peduncle. Nut is the real fruit and it is a drupe, kidney shaped and grey in colour. Nut varies in size, shape, weight and shelling percentage.

7.4 Importance of Cashew

Each and every part of cashew is useful to man. Leaves contain about 23% tannin which is used in tanning for dyeing fishing nets. Young leaves are sometimes used for diarrhoea and piles, also as tooth powder. Older and dry leaves form good mulch, when decomposed, serve as good manure. The bark contains high percentage of tannin (9-12%) which may be used for tanning, colouring or marking ink. A yellow cashew gum is found on the bark of the tree, which is obnoxious to insects. The fleshy stem is the source of refreshing beverages. A type of varnish can be produced from the fleshy stem. The reddish-brown wood is light which is used for packing cases, house posts, fencing poles, building boats and wood is used as charcoal. Cashew apple is famous for its high ascorbic acid content (Vitamin C) which is up to five times that of citrus fruit. It contains 10.15-12.5% sugars (mostly reducing) and about 0.35% acid (as malic). It is useful for digestion. Various products can be made from the apples, such as fruit juice, syrup canned fruit, candied fruit, pickles, jam, jelly, chutney, alcoholic drink, wine, spirit, alcohol and vinegar. Fenny (wine) is commonly prepared from cashew apple which is famous in Goa. The kernels are of high nutritive value. It is rich in protein, carbohydrate, unsaturated fats, minerals like calcium, phosphorous and iron and vitamins.

The shell contains a natural phenolic compound known as 'Cashewnut shell liquid'(CSNL) which is a valuable raw material for a number of polymer-based industries like paints and varnishes, resins, and decorative laminates, brake linings and rubber compounding resins.

Table 7.1: Composition of cashew apple and kernel

Sl.No.	Content	Cashew Apple (proportion in per cent or mg / 100 grams)	Cashew Kernel (in percentage)
1	Moisture	87.5 per cent	5.9 per cent
2	Carbohydrate	11.6 per cent	22.0 per cent
3	Protein	0.2 per cent	21.0 per cent
4	Fat	0.1 per cent	47.0 per cent
5	Tannic acid	0.5 mg/100 g	-
6	Minerals	0.2 per cent	-
7	Carotene	0.09	-
8	Vitamin C	0.26 mg/100g	-
9	Calcium	0.01 mg/100g	0.55 mg/100 g
10	Phosphorus	0.01 mg/100g	0.45 mg/100 g
11	Iron	0.2 mg/100g	5.0 mg/100 g

Source: Mandal (2000)

7.5 Climate and Soil

Cashew is a tropical plant and can thrive even at high temperatures. Young plants are sensitive to frost. The distribution of cashew is restricted to altitudes up to 700 m above mean sea level where the temperature does not fall below 20°C for prolonged period. Areas where the temperatures range from 20 to 30°C with an annual precipitation of 1000 - 2000 mm are ideal for cashew growing. However, temperatures above 36°C between the flowering and fruiting period could adversely affect the fruit setting and retention. Heavy rainfall, evenly distributed throughout the year is not favourable though the trees may grow and sometimes set fruit. Cashew needs a climate with a well-defined dry season of at least four months to produce the best yields. Coincidence of excessive rainfall and high relative humidity with flowering may result in flower/fruit drop and heavy incidence of fungal diseases.

The general notion is that "cashew is very modest in its soil requirements and can adapt itself to varying soil conditions without impairing productivity". While cashew can be grown in poor soils, its performance would be much better on good soils. The best soils for cashew are deep and well-drained sandy loams without a hard pan. Cashew also thrives on pure sandy soils, although mineral deficiencies are more likely to occur. Water stagnation and flooding are not congenial for cashew. Heavy clay soils with poor drainage and soils with pH more than 8.0 are not suitable for cashew cultivation. Excessive alkaline and saline soils also do not support its growth. Red sandy loam, lateritic soils and coastal sands with slightly acidic pH are best for cashew.

7.6 Choice of Cultivars

The cultivars of cashew nut suitable for different regions in India are presented in Table 7.2.

Table 7.2: Major cultivars of cashewnut in India

Sl. No.	State	Cultivars
1	Andhra Pradesh	BPP 1, BPP 2, BPP 3, BPP 4, BPP 4, BPP 5, BPP 6, EPM 9/8, T No. 39, T No. 1, T No. 56, m44/3, BPP 10, BPP 11
2	Karnataka	Ullal-1, Ullal-2, Ullal-3, Ullal-4, UN-50, NRCC 1, NRCC 3, Selection -1, Selection-2, Chintamani-1
3	Kerala	Kerala Anakkayam-1, BLA-39-4, K-30-1, K-22-1, NDR-2-1, BLA-139-1, BLA-273-1, M-25/1, M262-2, M3/4, Madakkathara-1, Madakkathara-2, Dhana, Priyanka
4	Maharashtra	Vengurla-1, Vengurla-2, Vengurla-3, Vengurla-4, Vengurla-5, Vengurla-6, Vengurla-7, Vengurla-8
5	Odisha	Bhubaneswar-1, WBDC-1
6	West Bengal	Jhargram-1
7	Tamil Nadu	Vridhachalam-1, Vridhachalam-2, Vridhachalam-3, VRI 4
8	Madhya Pradesh	T No. 40

Source: Bose, Mitra *et al.* (Ed) (1990), http://www.nrccashew.org

7.7 Propagation

Cashew is a cross-pollinated crop and exhibits wide variation in respect of nut, apple and yield of seedling progenies. Therefore, vegetative propagation has been advocated to mitigate this problem and for higher yield. However, seed is most important for raising seedlings for the use as rootstocks. The different propagation methods practised in cashew are as follows:

7.7.1 Air-Layering

Air layering is one of the most common methods of cashew propagation. Pencil thickness shoots of previous season's growth are selected and a ring of bark of 3 cm long is removed. The cinctured portion is applied with IBA 500 ppm in lanolin paste and a lump of well-soaked saw dust is placed and wrapped with polythene tape (25 × 25 cm). The ends of the tape are carefully tied and left for rooting. When roots are formed sufficiently (60-80 days), which are seen from polythene wrapper, the layers are severed and planted in pots. They are kept under partial shade till transplanting.

7.7.2 Softwood Grafting

Among the vegetative propagation methods soft wood grafting is more successful in most of the cashew growing areas. It is similar to epicotyl method of grafting but differ only with respect to age of the seedlings (root stocks) used for multiplication. In soft wood grafting 30-40 days old seedlings with 1 to 2 pairs of leaves are used as root stocks while grafting. However, usually soft portion of the seedlings at 15-20 cm from ground level is availed for grafting. For grafting 10-12 cm long pencil thick scion from current season growth should be selected and precured. Precuring is done by clipping off the lamina leaving the petiole intact on the shoots. Within few days these petioles drop-off indicating the shoots are getting cured. Due to storage of food material the shoots get thickened and the terminal bud appears swollen. This swollen condition indicates that shoots are ready for separation from the tree. Precuring is done to increase the meristematic activity in the axillary and terminal buds. Wedge technique is used for grafting. Two pairs of bottom leaves are retained on the stock and the stock is decapitated 5cm above the second pair of leaves and a vertical incision, along the length of the stump to 3.75cm from the top of the stock is made. The scion is prepared like a wedge. It is inserted into the stock and tied with polythene strip of 1.5 cm width and 30 cm length of 100 gauge. To create a humid atmosphere around the scion bud, polythene caps of 20 × 2.5 cm size of 100-gauge thickness is provided for 15-20 days till the buds sprout. The grafts are kept under shade or in a mist chamber preferably since humidity and temperature can be controlled. Application of NPK @ 150:20:100 ppm supplied through irrigation water helps in better survival of grafts. When the buds are sprouted, remove polythene caps and grafts are shifted to open place. The successful graft shows signs of growth within 3-4 weeks after grafting.

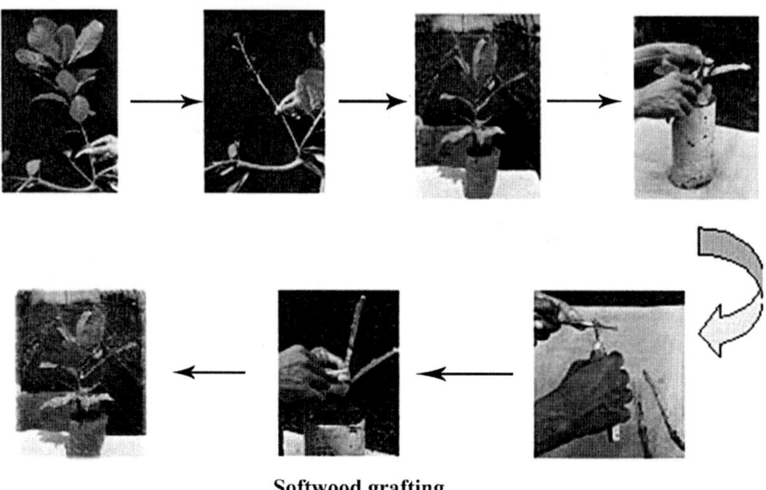

Softwood grafting

Source: TNAU Agritech.Portal, Horticulture *(See colour version on page 207)*

7.7.3 Top Working

Old and senile trees can be rejuvenated by top working/crown grafting with scions from superior genotype. Basically, the method of grafting is the same as described above for softwood grafting. Better management practices may increase the yields marginally but boosting cashew production 3-4 folds in a short span of time is perhaps possible only by "genetic transformation" of the existing plantations with high yielding cultivars. It is reported that this genetic transformation can be effected through top working. The rejuvenation of unthrift cashew plantations through top working involves beheading of trees, allowing juvenile shoots to start-out and taking up of in-situ grafting using procured scions of high yielding cultivars. Periods from November to March and February to June have been found to be ideal for beheading and in-situ grafting respectively. It has been observed that the top worked trees within a period of two years have not only put forth a canopy of 3-4 m in diameter and 5-6 m in height (as that of 8–10-year-old trees) but also have given a yield of 3 to 5 kg nuts per tree in their first bearing itself.

7.8 Preparation of Land

The land should be ploughed thoroughly and levelled in case of agricultural lands. In case of forestlands, the jungle should be cleared well in advance and the debris should be burnt. After clearing the jungles, land is to be terraced or bunds constructed on sloppy land. In order to ensure better moisture conservation, soil trenches are dug along the contours. The cost of land preparation will vary depending upon the type and method of soil working. Now-a-days, use of JCB for soil working is most popular; hence a provision for use of soil working is made in the model. The land preparation work should be completed prior to the onset of monsoon season, i.e., during May- June.

7.9 Layout

Cashew trees are generally planted with a spacing of 7 to 9 metres adopting square system. A spacing of 7.5 m × 7.5 m (175 plants/ha) or 8 m × 8 m (156 plants/ha) is recommended. High density planting of cashew at a closer spacing of 4 m × 4 m (625 plants/ha) in the beginning and thinning out in stages to maintain a final spacing of 8 m × 8 m in the tenth year is also recommended. This enables higher returns during initial years. In case of sloppy lands, the triangular system of planting is recommended to accommodate 15% more plants without affecting the growth and development of the trees. In undulating areas, the planting should preferably be done along the contours, with cradle pits or trenches provided at requisite spacing in a staggered manner to arrest soil erosion and help moisture conservation.

7.10 Digging and Filling of Pits

The work of digging of pits has to be completed much in advance (May-June). Cashew can be planted in pits of 60 cm × 60 cm × 60 cm size in soils with normal strata. In hard lateritic soils, pits of 1m × 1m × 1m size are recommended. The top soil and sub-soil are kept separately and allowed to wither under sun. It helps in migration of termites and ants. Burning of the debris and forest wastes inside the pits before planting is advantageous. The pits are then filled with topsoil mixed with farmyard manure or compost (5 kg) or poultry manure (2 kg) and rock phosphate (200 g).

7.11 Planting

The grafted plants obtained from the superior mother plant are usually planted at the onset of monsoon. It is essential to provide stakes and temporary shade with the locally available materials wherever necessary (especially in the South West aspects in case of forest plantation) to reduce the mortality rate and achieve quicker establishment. If the monsoon rains are inadequate, one or two pot irrigation can be done during the initial stages to ensure establishment.

7.12 Mulching

The cashew is generally planted on the wastelands and hence availability of soil moisture is always low. So, mulching is essential. Mulching with black polythene is beneficial to increase the growth and yield of cashew. However, locally available materials like green or dry grass or weeds can be utilized for mulching the basins. Small pebbles or stones can also be used for mulching of the basin. The plastic or stone mulch does not improve soil health but ensures better moisture retention in the soil and also prevents attack of soil borne insects and pests.

7.13 Nutrition

In our country, application of manures and fertilizers is very limited in the case of cashew. In order to get better yield, it is essential to maintain adequate N: P: K ratio in the soil. Application of 10-15 kg of farmyard manure per plant is recommended to ensure adequate organic matter in the soil. The fertilizers recommended for a mature cashew tree are 500 g N (1.1 kg urea), 125 g P_2O_5 (750 g Single Super Phosphate) and 125 g K_2O (200 g muriate of potash). The nutritional requirements and the quantity of fertilizer per plant are given in Table 7.3

Table 7.3: Nutritional requirements and recommended fertilizer doses for cashew

Age	Urea (g)	SSP (g)	MOP (g)
1st Year	375	275	75
2nd Year	750	525	150
3rd Year	1100	750	200

The ideal time for application of fertilizer is immediately after the cessation of heavy rains. Fertilisers should be applied in a circular trench along the drip line. Before application of fertilizer, it should be ensured that there is adequate soil moisture. The fertilizers should be applied in two split doses during pre-monsoon (May-June) and post-monsoon (September-October) season. However, in the case of single application, it should be done during post-monsoon season (September-October) when adequate soil moisture is available. In sandy and laterite soils, soils of sloppy land and in heavy rainfall zones, the fertilizer application should be done in a circular trench of 25 cm width and 15 cm depth at 1.5 m from the tree trunk. In red loamy soils and in low rainfall areas (east coast), the fertilizers should be applied in circular bands at a distance of 0.5 m, 0.7 m, 1.0 m and 1.5 m away from the trunk during first, second, third and fourth year onwards of planting, respectively.

7.14 Weeding

Weeding with a light digging should preferably be done before the end of rainy reason. Hoeing, cutting the weeds off underground is more effective than slashing. Chemical weeding has not been of any importance until now.

However, it may be considered as an alternative, where wages are high or where there is shortage of labour. Initially, Agrodar-96 (2, 4 –D) @ 4 ml/litre of water and subsequently Gramaxone @ 5 ml/litre of water is sprayed. Approximately, 400 litre/ha (160 litre/acre) of solution is required per spray. The spray is again repeated in the post-monsoon season.

7.15 Inter-Cropping

Tall growing intercrops like certain cultivars of sorghum and millet should not be encouraged between young cashew, as they provide too much shade. Leguminous crops such as groundnut and beans are very suitable for inter-cropping. Besides the annual crops, arid zone fruit crops having less canopy especially Annona, phalsa, etc., can be thought of, depending on the suitability. Cultivation of horse gram, cowpea, groundnut etc. is recommended as inter-crops in cashew. Inter cropping cashew with Casuarina and coconut is popular.

7.16 Cover Cropping

Leguminous cover crop enriches soil with the plant nutrients and adds organic matter, prevents soil erosion and conserves moisture. The seeds of these cover crops may be sown in the beginning of rainy season. The seed beds of 30 cm × 30 cm size are prepared in the interspace in slopes by loosening soil and mixing a little quantity of compost. The seeds of these crops are sown in the beds and covered with a thin layer of soil. The seeds should be soaked in the water for six hours before sowing.

7.17 Training and Pruning

During first year of planting, the sprouts coming from the rootstock should be removed frequently to ensure better health of the plant. These sprouts eat up valuable plant nutrition and also cause death of grafted scion allowing only rootstock to grow. Initial training and pruning of cashew plants during first 3-4 years is essential for providing proper shape to the trees. The trees are shaped by removing lower branches and water shoots coming from the base during first 3-4 years. Thereafter, little or no pruning is necessary. The plant should be allowed to grow by maintaining a single stem up to 0.75-1.0 m from the ground level. Weak and criss-cross branches are also chopped off. In order to avoid lodging of the plant by wind, proper staking of plant is essential. After 4-5 years, the main stem is detopped to a height of 4-5 m from the ground level. Thereafter, regular removal of dried/dead wood, criss-cross branches and water shoots once in 2-3 years is done to keep the plant healthy. The training and pruning of cashew plants is done during August- September. The cut surfaces are smeared with Bordeaux paste. The flowers appearing during first and second year of planting should be removed (de-blossoming) and plants should be allowed to bear fruits only after third year.

7.18 Crop Protection

7.18.1 Insect-Pests

It is observed that there are about 30 species of insects infesting cashew. Out of these tea mosquito, flower thrips, stem and root borer and fruit and nut borer are the major pests, which are reported to cause around 30% loss in yield.

1. Tea Mosquito

The nymphs and the adults of tea mosquito (*Helopeltis antoni*) suck sap on the tender leaves, shoots and inflorescence and even young nuts and apples. The saliva of the insect is very toxic, which causes blistering at the site of infestation. Severe attack on the young shoots cause dieback. Attacked inflorescence usually can be recognised from a distance by their scorched appearance. The affected panicles and shoots dry up in case of severe infestation. Though the pest is found throughout the year, the peak period of infestation is during the flushing, flowering and fruiting season (October-March). Tea mosquito can be controlled by spraying carbaryl 0.1% or phosalone 0.07% or dimethoate 0.05%. Spraying should be done thrice, first at the time of flushing, second at early flowering and third at the time of fruit set.

2. Stem and Root Borers (*Plocaederus ferrugineus*)

The young white grubs bore into the fresh tissues of the bark of the trunk and roots and feed on the subsequent subepidermal tissues and make tunnels in irregular directions. Due to severe damage to the vascular tissue the sap flow is arrested and the stem is weakened. The characteristic symptoms of damage include the presence of small holes in the collar region, gummosis, yellowing and shedding of the leaves and drying of the twigs. Once the plant is infested complete control of this pest is very difficult. However, prophylactic measures for its control can be adopted with 0.1% BHC swabbing twice a year, once in April-May and the second application during November.

3. Thrips

Both nymphs and adults suck and scrape at the underside of the leaves, mainly along main veins, causing yellowish patches, latter turning grey, giving the leaves a silvery appearance. The thrips are more active during the dry season.

Spraying of 0.05% monocrotophos or 0.1% carbaryl is very effective for controlling thrips.

4. Leaf and Blossom Webbers

The pest is assuming severe proportions in Tamil Nadu and Odisha. The symptoms of infestation include the presence of silken webs reinforced with pieces of plant parts on the terminal portions of tender shoots and blossoms, and the drying of webbed shoots. Spraying of 0.05% fenitrothion at the time of emergence of new flushes immediately after the monsoon is effective.

5. Leaf Miners (*Acrocercops syngramma*)

It is commonly observed in the post-harvest and post-monsoon flushes. Young trees are more prone to the attack of this pest. The silvery gray moths mine through tender leaves which later on swell up as blistered patches and severely damage them. Spraying of 0.05% fenitrothion or endosulphan 0.05% at the time of emergence of new flushes immediately after the monsoon is effective.

7.18.2 Diseases

1. Pink Disease or Die Back

The disease is caused by the fungus *Pellicularia salmonicolor* which is prevalent during the south-west monsoon. The affected branches initially show white patches on the bark and a film of silky thread of mycelium on the branches during monsoon. Later the fungus develops a pinkish growth which represents the spore mass and the affected shoots dry up from the tip downwards. Leaves develop into small sharply defined green spot with yellow background. Measures to check the spread of this disease include pruning of the affected branches or the shoots below the spot of infection and destroying them and applying Bordeaux paste (10%) on the cut surface to check further spread. Also spray 1% Bordeaux mixture twice, the first during May-June and the second in October as prophylactic measures.

2. Anthracnose (*Colletotrichum gloeosporioides*)

The fungus enters the fruit through the stigma in the flower stage. The attacked nuts exhibit small black spots and the apples become mummified. Rainy season is favourable for the spread of this disease. To control this disease, remove all the affected parts and spray the plant with 3:3:50 Bordeaux mixture and provide wind break by growing tall trees like *Casuarina*, *Eucalyptus* etc. to check the spread of disease by wind-blown spores.

3. Inflorescence Blight

The inflorescence blight is caused by fungi in association with tea mosquito. The disease is characterized by the drying of floral branches particularly when cloudy weather prevails. The mosquitos should be kept under control. Hence a combination spray of fungicide and insecticide is recommended to control this disease.

7.19 Harvesting and Yield

Cashew plants start bearing after three years of planting and reach full bearing during tenth year and continue giving remunerative yields for another 20 years. The cashew nuts are harvested during February-May. Normally, harvesting consists of picking of nuts that have dropped to the ground after maturing. However, if the apples are also used for making jam, juice, syrup, Fenni, etc., the fruit has to be harvested before it falls naturally. The cashew apples are removed and the nuts are dried in sun for 2-3 days to bring the moisture level from 25% to 9%. The maturity of the cashew nut is tested by floatation method. The mature nuts sink in water while the immature/ unfilled ones float. The nuts are collected at weekly intervals from the farm during the harvesting season. During that period the land should be clean in order to facilitate collection of cashew nuts. Plantations of unknown origin or seedling progenies with conventional methods of cultivation yield less than one kg of raw nuts per tree. However, there is a chance to increase the yield up to 4 to 5 kg per tree with the adoption of improved production techniques, over a period of 4 to 5 years. In new plantations, with the use of elite planting material coupled with a package of improved agronomic practices, a yield of 8-10 kg per tree could be achieved.

7.20 Cashew Processing

Processing of cashew is defined as the recovery of edible meat portion- the kernel from raw nuts, by manual or mechanical means. In India, the processing is done by manual means. It consists of: 1. roasting, 2. shelling, 3. peeling, 4. sweating, 5. grading and 6. packing.

Moisture Conditioning or Humidifying: A slight under-roasting or over-roasting adversely affects the quality of the kernels. This is achieved by a moisture step preceding the roasting. The raw nuts are sprinkled with water and allowed to remain in moist condition for about 24-48 hours. This step is known as conditioning. The optimum moisture level at the end of roasting is reported to be 15-20%. Two important points to be taken care of during conditioning are:

i. The water should not seep through the brown testa.

ii. The water should be free from iron contaminations. Iron contamination in the water can interact with polyphenolic materials of testa and the resultant bluish black complex may give patches on white kernel.

1. Roasting

Roasting is designed to make shell brittle. There are following types roasting:

(a) *Open Pan Roasting*: The earliest process was the pan roasting wherein the nuts are heated on a metal pan over an open fire. Due to the heat and slight charring the shells become brittle. The pan roasting is not followed in organized sectors of industries. The two important methods of processing now adopted are: drum roasting and oil bath roasting.

(b) *Drum Roasting*: The nuts are fed into a rotating hot drum, which ignites the shell portion of the nut. The drum maintains its temperature because of the oil oozing out of the nuts. The drum is kept in rotation by hand for about 2-4 minutes. The roasted nuts which are still burning are covered with wood ash to absorb the oil on the surface. The rate of shelling and the outturn of whole kernels are very high in this method. However, the main disadvantage is the loss of CNSL which has a very high export potential. In addition, there will be considerable heat and acrid fumes in the vicinity of this operation.

(c) *Oil Bath Roasting*: In this method, the nuts are held in wire trays and passed through a bath of cashew shell oil maintained at a temperature of 200-202° C for a period of three minutes whereby the shell oil is received from the shells to maximum possible extent. The vessel is embedded in brick work and heated by a furnace which uses spent shell as fuel. During roasting, the shell gets heated and cell wall gets separated releasing oil into bath. As the level rises the oil is recovered by continuous overflow arrangement. The roasted nuts are then converted into a centrifuge. The residual oil adhering to the surface of nuts is removed by centrifuging. The roasted nuts are mixed with wood ash and sent for shelling.

(d) In Panruti (Tamil Nadu) the conventional roasting is completely avoided. The raw nuts are exposed to the intense sun that is prevalent in that region. The well dried nuts are hand shelled. Here also the CNSL is completely recovered.

2. Shelling

After roasting, shelling is done by labour. Each nut is placed edgewise and cracked open with a light wooden mallet and the kernel extracted with or without wire prong. Care has to be taken that the inner kernel is intact and not broken into bits. After kernels are removed from the shells, they have to be dried to reduce the moisture to loosen the adhering testa.

3. Peeling

Peeling is the removal of testa from the kernels. This is done with help of safety pin or small hand knife. Peeling is made easier when the kernels are subjected to a heat treatment for about 4 hrs in a drying chamber.

4. Sweating

After peeling, the kernels are spread out indoors on cement flooring so that they may absorb some moisture and become less brittle. This prevents the tendency to break easily during grading.

5. Grading

The next stage in the processing is the grading of kernels on the basis of specifications for exportable grades. There are 25 exportable grades of cashew kernels. The kernels are stored into wholes, splits and broken primarily on the basis of visual characteristics. The wholes are again size-graded on the basis of the number of kernels per 1lb. The entire grading operation is done manually. However, for size-grading mechanical operation is also practiced.

Table 7.4: Specification for cashew kernels

Grade Designation	Number of kernels per lb	Grade Designation	Number of kernels per lb
W 180	375 to 395	W 320	660 to 705
W 210	440 to 465	W 400	770 to 880
W 240	485 to 530	W 450	880 to 990
W 280	575 to 620	W 500	990 to 1100

6. Packing

Final operation is packing in 10 kg capacity tins, which are subsequently evacuated and filled with carbon dioxide. In some parts to overcome the possible over-drying a rehumidification step is introduced before packing. The practice of filling with an inert gas is mainly to combat infestation during transit. It may be pointed out that with high quality nut, free from infestation, storing with or without carbon dioxide makes very little difference particularly with reference to rancidity. The importance of inert gas appears to be more for circumventing a possible insect attack from an occasional insect egg entering the tin while packing. Nitrogen can also do the same function. However, carbon dioxide being a heavier gas is more convenient for handling. Contention that absorption of carbon dioxide makes the kernel tastier does not have much truth. In any case the processed kernels are rarely consumed without a subsequent heat processing in the form of roasting frying and/or baking.

7.21 By-Products of Cashew

After the processing of the shell and other left outs are used for making some other products, the major by-products of cashew processing are: Cashew Nut Shell Liquid and Shell charcoal.

1. *Cashew Nut Shell Liquid* (CNSL): The pericarp of the nut consists of a coriaceous epicarp, spongy mesocarp and stony endocarp. The kernel covered with testa membrane is contained in a shell 1/8 inch thick. The mesocarp consists of a honeycomb network of cells containing a viscous liquid called cashew nut shell liquid (CNSL), which provides a natural protection to the kernel against insects. CNSL is a valuable raw material for a number of polymer-based industries like paints and varnishes, resins, industrial and decorative laminates, brake linings and rubber compounding resins. CNSL is traditionally obtained as a by-product during the isolation of kernel. The major constituents of shell oil are cardanol and anacardic acid of which cardanol is separately extracted and used in many industries. The shell oil was used as a preservative for boats and nets and to protect wood from termites. It is now largely exported and used in the manufacture of plastics, indelible inks, water proofing composition and other industrial products. The extraction of CNSL involves various methods, *viz.*, hot oil bath, expellers, kiln method, solvent extraction etc, the most common method being hot oil bath. In this method the raw nuts are passed through a bath of CNSL itself by which the CNSL is extracted. This method extracts only 50% of liquid contained in nuts. Then through expellers about 90% of liquid can be extracted.

2. *Cashew Shell Charcoal*: The remains of shell after the extraction of CNSL is called shell charcoal. This is used as a fuel. The shell charcoal is used in processing of cashew for drying after shelling

References

Bal JS 2006. *Fruit Growing*. Kalyani Publishers, Ludhiana, p. 153.

Bose TK, Mitra SK et al. (Ed) 1990. *Tropical Horticulture*, vol. 1. Naya Prokash, Calcutta pp. 566-572.

Gupta VK and Sharma SK 2000. *Disease of Fruit Crops*. Kalyani Publication, New Delhi.

Mandal RC 2000. *Cashew Production and Processing Technology*. Agrobios, India, p. 129.

Singh Amar 1996. *Fruit Physiology and Production*. Kalyani Publishers, Ludhiana, p. 486.

Chattopadhyay TK 2010. *A Textbook on Pomology (Tropical Fruits),* Vol-II. Kalyani Publisher, Ludhiana

Singh SP 2010. *Commercial Fruits*. Kalyani Publisher, Ludhiana

OUTCOME ASSESSMENTS

PART A

Answer the following questions true or false.

1. The cashew nuts are harvested during February-May. True/False
2. Mulching with black polythene is harmful to cashew. True/False
3. The major constituents of shell oil are cardanol and anacardic acid. True/ False
4. Groundnut and beans are not suitable for inter-cropping in cashew. True/ False
5. The inflorescence blight is caused by fungi in association with tea mosquito. True/False

PART B

Answer the following questions.

1. Why roasting of shell is required?
2. CNSL stands for —.
3. Cashew plants start bearing after three/six years of planting.
4. What is shell charcoal?
5. Cashew trees are generally planted with a spacing of — to — metres adopting square system.

PART C

Write a brief note on each of the following.

1. Propagation of cashew nut.
2. Roasting of shell.
3. Preparation of land and layout in cashew.
4. By-products of cashew.
5. Crop protection in cashewnut.

8

Production Technology of Coffee

Botanical Name: *Coffea arabica, Coffea canephora*
Family: Rubiaceae
Chromosome No: 44, 22
Place of Origin: Africa
Common Name: Coffee

8.1 Introduction

Coffee is widespread throughout the tropics with more than 70 species. All cultivated species originate from Africa. Economically important today is *Coffea arabica* (Arabica, 64 % of world production) and *Coffea canephora*, (Robusta, 35 %). It is cultivated worldwide on approximately 10.3 million hectares and represents the sole economic income for more than 25 million families. The crop is produced and exported by more than 60 nations and ranks as one of the top cash crops in developing countries with 3.19 lakh MT production per annum in India in 2018-19 from more than 459 thousand hectares crop area. The coffee plant is a fast-growing tropical bush tree, with two types of shoots: upright growing orthotropic main shoots (stem), and horizontally growing plagiotropic shoots (branches). Traditionally, coffee trees are cultivated only for the berries, which are processed using dry or wet techniques directly in the growing areas to the final raw product, green coffee; this serves as the basis for various coffee products. Both the coffee species tolerate shade and share quite similar growth requirements with various forest crops and trees, thus predisposing them to be grown in agro-forestry ecosystems.

8.2 Origin and Distribution

All commercial coffee species originated from Africa and belong to the genus *Coffea*. The high quality *Coffea arabica* species originates from the rainforests in the south western highlands of Ethiopia. One theory suggests that the Ethiopians took it to Yemen when they conquered the country by AD 500. Another hypothesis says that Arab merchants brought it initially to Yemen and the Arabian Peninsula, where it was cultivated and has contributed to the prosperity of the seaport of Mocca. This explains why Arabica coffee is associated with the name Mocca, although the prime centre of origin and diversity is on the African continent. Robusta coffee is resistant to coffee leaf rust (*Hemileia vastatrix*) and, therefore, with the expansion of coffee production in the world it replaced Arabica in the areas where coffee leaf rust was devastating the production. *Coffea liberica* originates from West Africa around Liberia. *Coffea excelsa* comes from the more continental and drier parts of Central Africa, mainly the Central African Republic. Nevertheless, practically all present cultivars are descendants of early coffee introductions from Ethiopia to Arabia (Yemen), where they were subjected to a relatively dry ecosystem without shade for a thousand years before being introduced to Asia and Latin America.

Now Brazil, Vietnam, Colombia, Indonesia, Ethiopia, Honduras, India, Mexico, Peru and Uganda are the top ten leading coffee producing countries in the world with Brazil as the highest global producer of coffee beans for over 150 years. Among all coffee producing states in India, Karnataka is the indisputable leader and produces more than 70% of the total coffee produced in the country with Kerala second position with more than 20%. Karnataka produced 2.33 Lakh Metric Tonnes of coffee in the last financial year, which is the highest coffee production by any state in India. Tamil Nadu, Andhra Pradesh, Odisha, Tripura, Nagaland, Assam, Meghalaya, Manipur are the other leading coffee producing states contributing rest 10%.

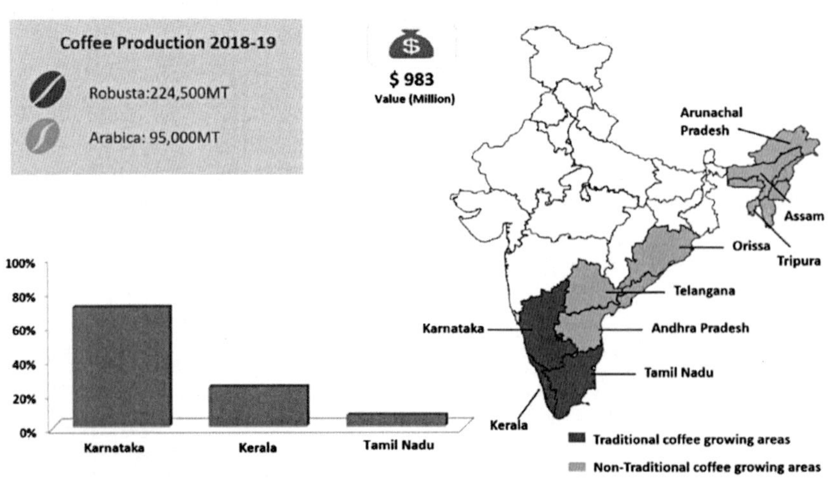

Main coffee producing states in India *(See colour version on page 207)*

8.3 Health Benefits of Coffee

- Coffee is a wonderful beverage that can reduce depression.
- Coffee is a natural source of antioxidants.
- Coffee can reduce the risk of diabetes (Type-2).
- Coffee is good for heart health and regular consumption of coffee may lower chances of heart disease and stroke.
- Coffee is a good agent which can prevent Alzheimer's disease. It can also boost a short-term memory.
- Coffee is an excellent drink for liver health. Regular consumption of coffee can reduce the chances of liver cells damage.
- Coffee is good at protecting against Parkinson's disease.
- Coffee may guard against the gout problem.
- Regular consumption of coffee may lower the risk of multiple sclerosis.
- Coffee can help in preventing a certain type of cancers such as colorectal cancer.

8.4 Botany

Cultivars and Classification

The genus *Coffea*, comprising more than 70 species, belongs to the family of Rubiaceae. This family forms part of the major group of dicotyledonous sympetales, wherein the petals of the flowers are fused. The coffee tree produces purple or red cherries (edible fruits) and these cherries can consist of seeds (coffee beans). Coffee trees are bush or tree type that can reach a height of 12 feet. The coffee plant flowers have an excellent aroma and usually exists in white colour. Once coffee orchard is established, it can produce coffee beans up to 55 to 60 years. Only three coffee species have commercial significance.

Coffee flowers
(See colour version on page 208)

Coffee beans

***Coffea arabica* Linné** – This species is divided in several varieties, some tall (Bourbon, Typica, …) and some dwarf (Caturra, Catuai, …). It is a tetraploid species (4x = 44) that yields a clearly superior coffee taste combining low caffeine content with fine aroma. It is generally susceptible to coffee leaf rust and, unfortunately, the more spread varieties like Bourbon tend to be the more susceptible. Among the more than 200 existing Arabica varieties, the most important tall varieties are:

Typica: Grown mainly in Brazil; most of the existing varieties originate from it

Bourbon: Has a 25% higher production than Typica

Mocha: Originates from Ethiopia

Mundo Novo: This is a natural cross between Bourbon and a variety from Sumatra

SL28 and Ruiru 11 from Kenya

Pache Comun and Pache Colis: Both are mutations from Typica

Maragogype: A mutation from Typica, characterized by broad beans; it originates from Brazil

Marella, Kent, S288 and S795: All of these originate from India

Blue Mountain: The famous variety from Jamaica.

The most important dwarf varieties are:

Caturra: A mutation from Bourbon, known for its productivity and good taste; originates from Brazil

Catuai: A cross between Caturra and Mundo Novo

Catimor: A cross between Caturra and Hybrido de Timor

***Coffea canephora* Pierre** – This is divided in many varieties, but two are mainly grown for commercial purposes: Robusta and Conilon (mainly grown in Brazil). As a whole, it is a diploid species (2n = 22) with an inferior taste but a higher yield. It is more as resistant against coffee leaf rust (*Hemileia vastatrix*), root-knot nematodes (*Meloidogyne exigua, M. incognita, M. paranaensis, Pratylenchus brachyurus,* and *P. coffeae*) and the coffee berry disease (*Colletotrichum kahawae)*. Important varieties are:

Robusta: The most common variety

Conilon: Grown mainly in Brazil

Kouilou or Kwilu: With smaller grains and fruits than Robusta

There exist also hybrids between *C. arabica* and *C. canephora*. Some are natural, as for example the Hybrido de Timor, and some are artificial, e.g., Arabusta created in Ivory Coast.

Table 8.1: Comparison between *C. arabica* and *C. canephora*

Sl. No.	Character	*C. arabica*	*C. canephora*
1	Ploidy	Tetra 2n=44	Diploid 2n=22
2	Plant Status	Small tree, Shrub/Bush	Bigger tree
3	Root	Small deep rooted	Large, Shallow
4	Branches	Persistent	Deciduous after harvest
5	Leaves	Dark green	Pale green
6	Flowering part	On new wood	On new and old wood
7	Regularity in bearing	Regular	Irregular
8	Flowers	Scaly, Small bracts,4-5 inflorescence, at each node.1-4 flowers	Leaf expanded bracts, 5-6 flowers/inflorescence
9	Pollination	Self-fertile, Self-pollinated	Self-sterile, Cross-pollinated
10	Berries	Medium in size, 10-20/node	Small, 40-60/node
11	Fruit development	8-9 months	10-11 months
12	Caffeine content	1.47 %	2.2 %

***Coffea liberica* Hiern** – This is a diploid species (2n = 22) which includes several varieties like var. liberica, var. dewevrei and var. excelsa. It is known for its pungent and earthy tasting coffee. Though it combines good rust resistance, the species is declining and has become less significant as an international commodity.

8.5 Climate and Soil

Climatic conditions play a major role in coffee growing. The Coffee Arabica can be cultivated at high altitude from 900 meters to 1600 meters. However, it can be grown well to lower elevations as well. The main disadvantage of cultivating coffee beans at high altitudes is that it can result in late maturity and the crop may be damaged due to frost conditions and high winds. When it comes to rainfall requirement, the coffee crop grows well with an evenly distributed annual rainfall of 2200 to 2300 mm. This crop is very sensitive to high rainfall conditions and this may cause rotting of leaves and fungal disease. It is important that the rains are well distributed over the season or are continuous for about 7-8 months. The ideal temperature range is from 18 to 28° C. High temperatures above 30°C during blossoming associated with a prolonged dry season may cause abortion of flowers. Coffee plantation requires shade in regions where high temperatures are possible. Strong winds can reduce the leaf area and the inter-nodal length of the orthotropic and plagiotropic branches and hamper the development of flowers and fruits.

Hot winds increase crop evapo-transpiration, thus increasing the moisture requirements of the trees. Where strong winds are frequent, windbreaks or shelter trees are recommended to improve crop performance. Growing coffee under full sun with little or no forest canopy causes the berries to ripen more rapidly and the bushes to produce higher yields; this requires shade clearing and increased fertilizer and pesticide uses. In traditional coffee production, however, with different levels of shade, berries ripen more slowly, and yields are lower, but green coffee and cup quality are superior.

S.No.	Climatic factors	Arabica	Robusta
1	Climate	Temperate climate in tropics	Warm humid conditions
2	Elevation	900 – 1500 m	500 – 1000 m
3	Annual R F	1600 – 3000 mm (over 8-9)	1000 –2000 mm
4	Temp	15 –35 ^{0}C	20 –30^{0}C
5	RH	70 –80 %	80 –90%
6	Shade	Medium to light shade	Uniform thin shade
7	Blossom rain	Mar – Apr	Feb – Mar

Coffee can be grown on hill slopes and on lands with undulating topography. The soil should be deep, friable, porous in nature, with good aeration, rich in organic matter, with good water holding capacity, rich in potassium and slightly acidic (pH 6.0 – 6.5). In India, coffee is cultivated in red and lateritic soils.

8.6 Propagation

Coffee is propagated both by seed and vegetative (cuttings) method. The plants that are good shaped, high and regular yielder, disease resistant, and with excellent quality beans are marked as mother plants. Healthy, fully mature, ripe fruits are collected. Hand pulped with ash and extracted seeds are dried in shade. Seeds are sown within two months of gathering or seeds are stored with dry powdered charcoal for 10 months without loss in viability. Select light to medium loam, humus rich, nematode free, pest free, gentle slope, partially sheltered site for nursery.

Raised beds (15 cm) of one metre width and convenient length are prepared. Four baskets of FYM/ compost, 2 kg lime and 500 g rock phosphate are incorporated into a bed of 1 m × 6 m size. Seed treated with agrosan is sown in December-January months flat side facing the soil. Seeds are sown 2.5 cm apart in the rows, which are spaced at 15 – 20 cm, covered with thin layer of fine soil and bed is covered with 5 cm paddy straw to ensure even temperature and conserve moisture. Beds are watered daily, protected from direct sun light by erecting an overhead pandal. Germination commences in four weeks and will be completed within another 5 to 6 weeks. When the seedlings are 5–8 cm tall, they are shifted either to secondary nursery beds or to poly bags of 150 gauge during Feb –March. Secondary nursery beds are prepared in the same way as in the case of seed beds. Seedlings are transplanted at a spacing of 15 –20 cm within the rows spaced at 30 cm apart. Regular watering and after care of the seedlings should follow. Overhead shade is gradually thinned and removed before the onset of monsoon. Seedlings are maintained in the secondary nursery beds for 16–18 months. Seedlings are manured once in two months with urea solution @ 20 g in 4.5 litres of water.

Cuttings: Cuttings are obtained from bushes of outstanding performance are free from pests and diseases. Single node, 3–6 m old, semihard wood (greenwood) cuttings of 1.0 cm long with two basal leaves are prepared. To enhance rooting, the bases are dipped in 5000 ppm BA at planting. Cuttings are planted in poly bags filled with forest soil, sand and FYM during rainy season (June, August). Poly bags with cuttings are arranged in a trench (2 m width, 50 cm depth and of convenient length) covered with thick polythene sheet over a framework of bamboos or aluminium frame. Cuttings will root in about 3 months after planting and then they are hardened under shade for 2 months. Propagation by cuttings is not an established practice in India.

8.7 Land Preparation, Transplanting and Spacing

For better coffee orchard establishment, the land should be prepared very well by giving 4 or 5 ploughings and harrowing to bring the soil to fine tilth. As part of this preparation, remove any stones, debris and weeds from previous crops. Commercial coffee growers should consider soil testing to measure soil fertility and suitability. Based on soil test reports, any nutrients and micronutrients should be supplemented in the soil before planting the seedlings. In the case of fertilizer applications such as phosphorus and lime, these should be thoroughly incorporated by ploughing and disking the soil several months before transplanting the coffee seedlings in the field.

Planting distance or space vary from variety to variety and mainly depends on topography and soil fertility. The general spacing of different species of coffee is as follows:

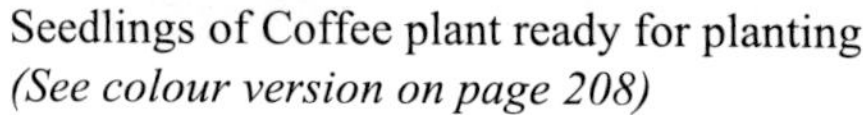

Seedlings of Coffee plant ready for planting
(See colour version on page 208)

Coffee plantation in India

1. Coffee Arabica: 2 m × 3 m
2. Coffee Robusta: 3 m × 3 m

When it comes to the plantation, straight row planting with an east-west orientation is preferred layout in coffee growing. The coffee crop is planted during the spring season or just before the rainy season as this crop requires moist soil conditions and cool climate for better establishment and growth. Six to seven months old seedlings of about 20 to 25 cm height should be transplanted in the main field. Prepare the pits in advance to loosen up the soil before transplanting the coffee seedlings in the field. These pits should be dug 3 months before transplanting the seedlings in the field. The pits should be dug with 50 cm × 50 cm × 50 cm size. In regions where sunlight is intense, this crop requires partial shade; for this purpose, is advised growing shade trees in the field. One can plant species of *Ficus*, *Terminalia* sp., silver oak, *Albizia lebbeck*, *Erythrina lithosperma*, *Artocarpus integrifoila* and *Gliricidia maculata* for the purpose of providing shade in coffee growing.

8.8 Water Management

The frequency of irrigation in coffee growing depends on the type of soil, moisture level in the soil, plant age and climate. There are many irrigation methods available like drip irrigation, sprinkler irrigation, microjet and basin irrigation that can be adopted in coffee plantations. However, drip irrigation is the best method for best utilization of water and fertilizers. This can also control weed growth at plant basins. Under irrigated conditions, each coffee plant requires 50 to 55 mm water and this should be applied before transplanting. Subsequent irrigations of 25 mm should be provided at 8 to 10 days interval. Make sure to keep the soil moist and avoid too much wetness.

Overwatering can result in plant rot and fungal diseases in coffee plantation. The thumb rule is to provide two irrigations in a week in dry climatic conditions (like in summer) and one irrigation per week in the cool winter season.

8.9 Nutrition

Timely application of fertilizers and manures ensures the high yield of coffee beans. The requirement of manures and fertilizers in coffee growing is:

Coffee Species	Pre-Blossom March month N: P_2O_5: K_2O	Post-blossom May month N: P_2O_5: K_2O	Mid-monsoon August month N: P_2O_5: K_2O	Post-monsoon October month N: P_2O_5: K_2O	Total quantity
Coffea arabica					
Young coffee 1st year after planting	15:10:15	15:10:15	—	15:10:15	45:30:45
Second and Third year	20:10:20	20:10:20	—	20:15:20	60:45:60
Fourth year	30:20:30	20:20:20	—	30:20:30	80:60:80
Bearing coffee 5 years and above for less than 1 tonne per ha crop	40:30:40	40:30:40	—	40:30:40	140:90:120
For 1 tonne per ha and above	40:30:40	40:30:40	40:30:40	40:30:40	160:120:160
Coffea robusta					
For less than 1 tonne per ha crop	40:30:40	—	—	40:30:40	80:60:80
For 1 tonne per ha and above	40:30:40	40:30:40	—	40:30:40	120:90:120

8.10 Intercultural Operation

Weed Control: Digging and forking should be done during the months of September, November and again in February and March. Weed control can be done during the initial years. Chemical weedicides like Dalapon can be used for controlling grasses. Amino salt of 2, 4 – D can be used for controlling broad-leaved weeds.

Mulching: Mulching can control weed, and prevent moisture loss and soil erosion. It can enrich the soil, moderate excessive soil temperature and improve soil texture. In most of the cases, mulch materials can be converted into excellent organic matter.

8.11 Training and Pruning

Training of the bush is necessary to have a strong framework which promotes producing of bearing wood. Coffee is trained in two systems:

1. *Single Stem System*: When the plant reaches a height of 75 cm in Arabica or 110 to 120 cm in Robusta, tree is topped. This helps to restrict vertical growth, facilitate lateral spreading and increase the bearing area. In this system, a second tier is also allowed sometimes depending upon the soil fertility and plant's vigour.
2. *Multiple Stem System*: It is common in Kenya, Tanzania, but not practiced in India. Pruning in coffee is generally done immediately after harvest and till the onset of monsoon. It is essentially a thinning process and is done mainly to divert the vigour of the plants to certain parts by pruning the other parts. Pruning involves:
    a. Centering: Removal of the vegetative growth up to 15 cm radius from the centre and up to the first node of all primary branches,
    b. Desuckering: Removal of small sprouts arising from the axils of the leaves which otherwise grow towards the inner side and cause shade and become unproductive wood and
    c. Nipping: Growing tip of primary branches is removed to encourage secondaries and tertiaries.

8.12 Soil Management

Soil management practices aim at conserving soil and water and in general to make the soil perform its functions satisfactorily. It includes the following practices in coffee:

a. *Digging*: In the new clearing, the field is given a thorough digging to a depth of about 35 to 45 cm towards the end of the monsoon. All weeds and vegetative debris are completely turned under and buried in the soil while the stumps are removed. Once the coffee plants have closed in, annual digging is not done.

b. *Scuffing or soil stirring*: In established coffee fields, scuffing or soil stirring is done towards the beginning of the dry period. It controls weeds and also conserves soil moisture.

c. *Trenching*: Trenches and pits are dug or renovated in a staggered manner between rows of coffee along the contour during August-October when the soil is fairly easy to work. These are 50 cm wide and 25 cm deep and can be of any convenient length.

d. *Mulching*: Mulching young coffee clearings helps to maintain optimum soil temperature and conserve soil moisture and acts as an effective erosion control measure. Mulching also adds to fertility of the soil.

e. *Weed control*: New clearings are hand-weeded three to four times a year and established coffee two to three times. During the monsoon, the weeds are slashed back. Another weeding is done towards the end of the monsoon. Clean weeding is generally done during the post-monsoon period. Chemical weedicides have gained popularity in larger plantations. Gramoxone at 1.25 litres in 450 litres of water per hectare has been found to be the best. This should follow slash weeded plots after 10-15 days.

f. *Irrigation*: Sprinkler irrigation is mainly used as an insurance against failure of good blossom or backing showers. It is also used on young plantations, marginal areas where water is available in plenty to help in establishment of coffee and shade.

g. *Soil acidity and liming*: The heavy rainfall in coffee growing zones of South India brings about leaching of calcium and magnesium leading to soil acidity. Besides, continuous use of acid forming fertilizers like ammonium sulphate also make the soil acidic. As the ill effects of soil acidity are more, periodical lime application is essential to correct the soil pH for good productivity. The quantity of lime to be applied is based on soil pH and lime requirement. Agricultural lime and dolomitic lime are the most commonly used liming materials. It can be applied to the soil at any time during the year provided there is a gap of one month or a few showers between lime and fertilizer applications. It is also desirable to apply lime when there is sufficient moisture in the soil for quick response.

8.13 Shade Management

Under the climatic conditions existing in India, coffee is being cultivated under shade. It comprises two canopies — lower or temporary and upper or permanent. High light intensities and temperature prevailing during the drought period are not conducive for normal and healthy growth of coffee plants. Therefore, there is necessity for protecting the coffee plants during the above period by providing both temporary (lower) and permanent (upper) shade trees.

Dadap (*Erythrina lithosperma*) is used as a lower canopy shade in India. It is planted along with coffee in new clearings. When stakes are planted in June, they grow quickly using the moisture available in the soil. Next to dadap, silver oak (*Grevillea robusta*) is the most commonly used tree for temporary shade. The most popular permanent shade trees found in South India are *Albizzia lebbek, A. moluccana, A.odoratissima, Artocarpus integrifolia, Cedrella toona, Dalbergia latifolia, Ficus glomerata, F. infectoria, F. retusa* and *Maesopsis eminii.*

Planting shade trees in Coffee plantation *(See colour version on page 208)*

Permanent shade trees are generally planted about 12 to 14 metres apart. It is advisable to plant a large number initially and thin out as the trees grow and spread out. The trees have to be regulated in such a way that in course of time, they have their canopy about 10 to 14 metres above the coffee. Shade trees require constant attention by way of pruning and lopping to provide the required filtered shade to coffee. The most convenient time to regulate shade is after pruning and liming.

8.14 Crop Protection

8.14.1 Insect-Pests

Insects generally weaken coffee beans and reduce density. The bites from insects also open coffee plants up to secondary infection from fungi and other microorganisms. Infestation by insects not only reduces yield, but also can have a considerable effect on coffee profile, with reduction in quality of flavour and aroma.

These are some of the important pests seen in a coffee farm.

1. Coffee Berry Borer *(Hypothenemus hampei)*

These tiny black beetles are present in almost all coffee-producing countries, where they burrow inside coffee cherries. They are very difficult to manage with insecticides because they are protected by the cherries. The insects spread worldwide from Africa alongside coffee crops as far back as the 16th century. Café de Colombia states that this pest has caused the most damage to coffee throughout history.

In crops affected by coffee borer beetles, yields are reduced because young bored cherries may fall prematurely and all harvested bored cherries are of lower weight. Coffee berry borer damage also affects the sensory qualities of the coffee and this reduces the commercial value of the crop. If damage is significant, it can cause the cup to taste bitter, tarry, or fermented. Berry borer damage can also cause roasts to be irregular, which has a further impact on flavour.

Management

Clean harvest, no berries should be left over either on plant or on ground. Maintain optimum shade. Dry coffee to prescribed moisture levels.

2. Coffee Leaf Miner

Coffee leaf miners are two related species of moth – *Leucoptera coffeella*, which is prevalent in Latin America, and *Leucoptera caffeina*, which is found in African producing countries. They affect the leaves of the coffee plant. Café de Colombia explained that coffee leaf miner larvae eat coffee leaves. If several larvae live on the same leaf, it may suffer necrosis of up to 90% of its structure. Necrosis is the death of cells and it appears as dark watery spots or brown papery patches. Defoliation affects the plant's ability to photosynthesize. Without photosynthesis, the plant cannot grow properly. Fruit may not mature and the overall yield is likely to be much lower. If immature or dead beans make it into the final brew, they can create bitterness and astringency.

3. Mealy Bugs (*Planococcus citri*)

Mealy bugs are a group of insects that feed on a variety of trees and plants. In coffee, they attack various parts including branches, nodes, leaves, roots and flower clusters. They feed on the sap of the coffee plant and secrete a sticky substance that attracts ants. This substance also leads to the formation of a black mould that covers leaves and can reduce photosynthesis.

Reduced sap uptake, circulation and photosynthesis stress coffee plants and they tend to produce light or immature beans. This can result in astringency, a metallic taste, or bitterness in the cup. Coffee mealy bugs have been found in Africa, Australia, Asia, and Central and South America.

Management

Maintain optimum shade. Control ants by dusting Malathion 5% in the base of shade trees. Destroy ant nests, remove and destroy weeds. Spray affected patches with Fenitrothion 50 EC @ 300 ml each or 4 litres of kerosene in 200 litres of water along with 200 ml of wetting agent. Release parasitoid *Leptomastix dactylopii*.

8.14.2 Nematodes (*Pratylenchus coffeae*)

Nematodes are microscopic worm-like parasites. There are several species that attack the root system of coffee plants and feed on their sap. Nematodes can form knots in the roots that prevent the plant from properly absorbing water and nutrients. Infestation can cause reduced roots, defoliation and general lack of health in the plants. This can mean low yield and light beans.

Management

Uproot and burn the affected plants. Dig up the pits and expose the soil to the sun for one year. Take care to keep the pits free from weeds. Plant the area with Robusta or Arabica- Robusta grafted plants (Arabica scion grafted on to Robusta rootstock).

8.14.3 Diseases

Disease is also a big threat to coffee production. These are some common diseases.

1. Coffee Leaf Rust (*Hemileia vastatrix*)

This disease is a worldwide problem for coffee producers and Colombia has been battling it for generations. The disease presents an orange rust-like dust on the underside of the coffee leaves. It is a cyclical condition that causes defoliation, just like coffee leaf miners. Wind and rain spread coffee leaf rust spores, which thrive at around 70°F/21°C. So, the disease is most prevalent in Arabica grown in the warm, humid conditions of low altitudes. Because it restricts the growth of new stems, coffee leaf rust has an impact on the next year's crop as well as significantly reducing yield in the current year. Plants affected by coffee leaf rust are unable to ripen fully and if they do, fruit will produce light beans that taste astringent.

Management

Three sprays of Bordeaux mixture 0.5% (1 kg $CuSO_4$: 1kg CaO : 200 litres of water), i.e., pre-blossom (February-March), pre-monsoon (May-June) and post-monsoon (September-October) or 3 sprays of Pantvax 0.3% (300ml/200 litre water), i.e., pre-monsoon (May-June), mid-monsoon (July-August), post-monsoon (September-October).

2. Coffee Wilt

Coffee wilt is a vascular disease of the coffee tree trunk that is caused by a fungus. It blocks water and sap circulation, causing leaves to fall, branches to die, and cherries to appear ripe prematurely. Using these red but immature coffee cherries may result in loss of acidity, increased bitterness, and "green" flavour in the cup.

3. Pink Disease

Pink disease is another fungal infection. It appears as webbing and pink encrustation on branches. The infected branches lose their leaves and die.

8.15 Harvesting and Processing

Coffee beans/cherries start bearing from the third year onwards after transplanting in the field. Healthy and best yield can be obtained from the fifth year. The best part is, the coffee plantation continues to yield up to 50 to 55 years. Coffee fruits should be picked as and when they become ripe to get better quality. Arabica comes for harvesting earlier since they take 8-9 months for fruit development from flowering while Robusta takes 10-11 months. Picking is done by hand. The picking consists of as follows:

- **Fly picking:** You can pick ripe berries of the coffee plant during the month of October to February. Generally, this is a small-scale picking.
- **Main picking:** It is recommended to pick well-ripened and formed coffee berries/beans during the month of December. This is the main harvesting method of coffee berries (used for bulk yields).

- **Stripping:** This method is used for picking all the berries left on the plant irrespective of the ripening stage.
- **Cleanings:** This method involves collecting the fruits that have been dropped during harvesting.

Coffee is processed in two ways: a) wet processing to prepare parchment coffee and b) dry method by which cherry coffee is prepared.

A) Preparation of parchment coffee

i. **Pulping**: This method requires pulping equipment and adequate supply of clean water. Fruits should be pulped on the same day to avoid fermentation before pulping. Fruits may be fed to the pulper through siphon arrangements to ensure uniform feeding and to separate lights and floats from sound fruits. The pulped parchment should be sieved to eliminate any unpulped fruits and fruit skin. The skins separated by pulping should be let away from the vats into collection pits so that microbial decomposition of the skin will not affect the bean quality when it gets mixed up with the bean.

ii. **Demucilaging and washing**: The mucilage on the parchment skin can be removed by:

   a. **Natural fermentation**: The mucilage breaks down in the process of fermentation and it takes 24 to 36 hours for Arabica and 72 hours for Robusta. Cool weather delays the process of fermentation. Under-fermented or over-fermented beans affect quality. When correctly fermented, the mucilage comes off easily and the parchment does not stick to the hand after washing and the beans feel rough and gritty when squeezed by hand. When the mucilage breakdown is complete, clean water is let in and the parchment is washed clean with three to four changes of water.

   b. **Treatment with alkali**: Removal of mucilage by treatment with alkali takes about one hour for Arabica and one and a half to two hours for Robusta. The beans obtained after pulping are drained off excess water and spread out in the vats uniformly and furrowed with wooden ladles with long handles. A 10% solution of caustic soda (sodium hydroxide) is evenly applied into the furrows using a water can. Ten litres of the alkali is sufficient to treat 25 to 30 forlits (one forlit = 40 litres) of parchment. The parchment is agitated thoroughly by the ladles so as to make the alkali to come into contact with the parchment and trampled by feet for about half an hour. When the parchment is no longer slimy and makes a rattling noise, clean water is let in and the parchment is washed clean with three or four changes of water.

c. **Removal of mucilage by friction**: There are machines, which pulp and demucilage the beans in one operation. However, a number of naked and bruised beans may result in the parchment. It is, therefore, necessary to adjust the machines carefully to obtain uniform pulping and demucilaging.

iii) **Drying**: The next stage is drying the parchment in the sun until the moisture content is sufficiently reduced to permit storage of beans till they are dispatched to curing works. Proper drying contributes to the healthy colour of the bean and other quality factors. Under-dried parchment turns mouldy and gets bleached during storage and subsequent curing operations.

The parchment is spread on clean tiled or concrete drying floor to be dried slowly by spreading to a thickness of about 7 to 10 cm. Stirring and turning over coffee, at least once an hour, is necessary to facilitate uniform drying. The parchment should be heaped up and covered in the evening until next morning. Sun drying may take about 7 to 10 days under bright weather conditions. At the right stage of dryness, the parchment becomes crumbly and the beans split clean without a white fracture when bitten between the teeth. Drying is complete when a sample forlit of coffee records the same weight for two days consecutively. At this stage, coffee is shifted to the stores and bagged in clean, new gunnies.

When coffee is being dried, all naked beans, pulper nipped and bruised beans, blacks, greens and other defective beans are sorted out and dispatched to curing works separately.

B) Preparation of cherry coffee

For preparation of cherry coffee, fruits should be picked as and when they ripe. Greens and under-ripe fruits should be sorted out and dried separately. The fruits should be spread evenly to a thickness of about 8 cm on clean drying ground in which the cherries are stirred and ridged at least once every hour. The cherry is dry when a fistful of the drying cherry produces a rattling sound when shaken and a sample forlit records the same weight on two consecutive days. The cherry should be fully dry at the end of 12 to 15 days under bright weather conditions.

References

Bansil PC 2015. *Plantation crops*. CBS Publishers & Distributors, New Delhi.

Hermann A. Jürgen Pohlan, Marc J. J. Janssens 2010. *Soils, Plant Growth and Crop Production* – Vo. III - Growth and Production of Coffee. Eolss Publishers, Oxford, UK, Editors: Ed. Willy H. Verheye

ICAR 2010. *Handbook of Horticulture*. ICAR, New Delhi.

Kumar N 2018. *Introduction to Spices, Plantation Crops, Medicinal and Aromatic Plants* (2nd Ed). Oxford & IBH Publishing Co Pvt. Ltd., New Delhi.

e-re*Source*s:

www.agrifarming.in> coffee-growing-information-beg

www.eolss.net>sample-chapter

www.agritech.tnau.ac.in>horticulture > hortiplantation

www.indiaagronet.com > horticulture

Perfectdailygrind.com > farming

OUTCOME ASSESSMENTS

PART A

Answer the following questions true or false.

1. All the commercial coffee species originated from South America. True/False
2. Coffee starts bearing from the third year onwards after transplanting. True/False
3. Coffee is processed in three ways. True/False
4. In India coffee is cultivated under shade. True/False
5. Climatic conditions play a minor role in coffee growing. True/False

PART B

Answer the following questions.

1. Coffee is propagated both by — and — methods.
2. Coffee is widespread throughout the tropics with more than 70/80 species.
3. What is pulping?
4. What is the centre of origin of coffee?
5. Explain demucilaging.

PART C

Write a brief note on each of the following.

1. Shade management in coffee.
2. Harvesting and processing of coffee.
3. Soil management in coffee.
4. Coffee berry borer.
5. Preparation of parchment coffee.

9

Production Technology of Tea

Botanical Name: *Camellia sinensis* L.
Family: Theaceae
Chromosome No: 30
Place of Origin: China
Common Name: Tea

9.1 Introduction

Tea (*Camellia sinensis* L.) of family Camelliaceae or Theaceae is the oldest and second most widely consumed beverage in the world following water. Tea is an evergreen plant that mainly grows in tropical and subtropical climates. India is world's 2nd largest tea producer after China with production of 1.25 million tonnes in 2020. In India it is grown in an area of 6 lakh hectares, contributing one fourth of world total tea production annually. Interestingly, India is also the world's largest consumer of the beverage and using nearly 30 percent of the world's tea output. India stands third in terms of tea export after Kenya (including neighbouring African countries) and China. The total tea export was US$ 501.16 million in April 2020 to November 2020 and for November 2020 it was US$ 71.61 million.

Fascinating aroma, attractive taste and many health effects of tea make it a popular beverage worldwide. All tea types belong to the same family of Camelliaceae. Application of various manufacturing process on fresh leaves of *camellia sinensis* produces white, green, black, yellow or oolong tea which their brewers differ in colour and taste. Polyphenols, caffeine and theanine are the main bioactive ingredients in *camellia sinensis*. Tea has a stimulating effect in human due to its caffeine content.

Difference between green tea and black tea

	Green Tea	Black Tea
Process	Short, no fermentation	Longer Fermentation
Colour	Green or Yellow	Red or Black
Taste	Sweet after bitter	Distinct flavour, added in sugar and/ or milk
Antioxidant	In general, more polyphenols	More flavonoids
Caffeine	Less	More
Quality	Better in fresh	Depend on produced location
Health benefits	More in general: may irritate to empty stomach	Cardiovascular system

Source: Sharangi, 2009.

Area and distribution

Tea is the native of China in South East Asia. It was known to the Chinese as early as 2737 BC, but attained the status of a popular drink in England in 1664 AD. In May 1838 the first Indian tea from Assam was sent to England for public sale. The other leading producers are China, Kenya, Sri Lanka, Turkey and Vietnam. Tea is grown in 16 states in India. Major tea growing areas of India are Assam (53%), West Bengal (23%), Tamil Nadu (11%) and Kerala (8%).

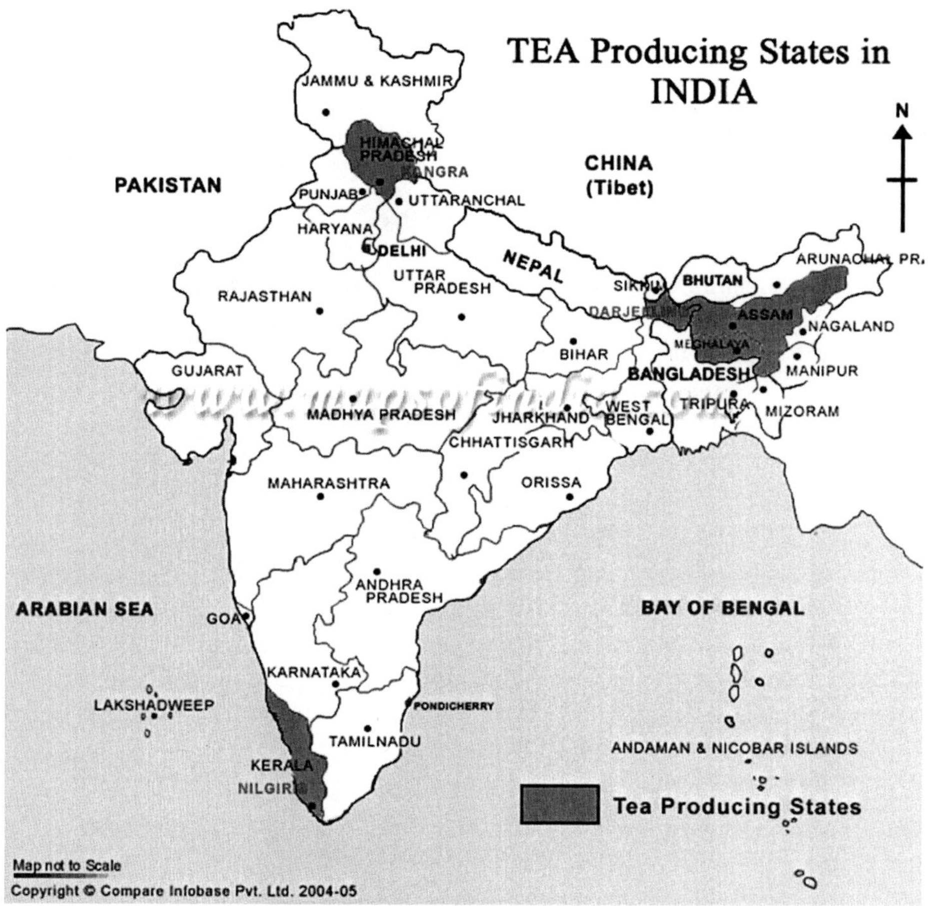

(See colour version on page 209)

9.2 Botany

Morphologically, tea is an evergreen shrub or tree, leaves are simple, alternate, serrate, flowers are bisexual, with superior ovary, fruit is a capsule. Presently three types of tea have been recognized as follows:

i) China tea plant – *Camellia sinensis* L.

ii) Assam tea plant – *Camellia assamica* (Masters)

iii) Cambod or Southern form of tea – *Camellia assamica* ssp. *lasiocalyx*

Being a highly cross-pollinated crop, the present-day seedling populations are mixture of both the above two species.

Morphology of Various Types of Tea

i) Assam Tea Plant

Assam tea plant under natural condition is a small tree which is about 10-15 m tall with a trunk sometimes up to one third of its height. It possesses a robust branch system. In typical plants, the leaf is droopy, thin, glossy or shining with more or less acuminate apex and distinct marginal veins. Leaf blades are usually 8-20 cm long and 3.5 -7.5 cm wide and light green to dark green in colour.

(See colour version on page 210)

ii) China Tea Plant

It is a big shrub; 1-3 m tall with many virgate stems. Virgate stem means straight or slender stem arising from the base of the tree near ground level. The leaf is hard, thick and leathery. The surface is mat (dull surface which is not shining), marginal veins are indistinct. The leaf blades of China tea plants are dark green in colour. Petioles are short, 3-7 mm long, stout, usually giving the leaf an erect pose.

iii) Cambod Tea Plant

Two third of Cambod tea plant is a small tree 6-10 m tall with more or less equally developed ascending main stems. It is by habit a fastigiate tree. Fastigiate means strictly erect and more or less parallel. The leaves of this species of tea plant are more or less erect, glossy, light green in colour, occasionally pigmented with anthocyanin. In size, the leaves are intermediate between China and Assam tea plant leaves.

(See colour version on page 210)

9.3 Cultivars

Major cultivars of tea: Pandian, Sundaram, Golconda, Jayaram, Evergreen, Athrey, Brookeland, BSS 1, BSS 2, BSS 3, BSS 4, BSS 5, Biclonal seed stocks and Grafts.

Clonal selection from seedling population was taken up by UPASI, Tea Scientific Department, Cinchona and also by other Tea Research Institutes. UPASI has so far released 27 clones. Certain outstanding clones released by other Institutes are also used in South India.

Tea cultivars with special characters			
S. No	Special characters	Clone	Originators
1	Wind Tolerance	UPASI-2, UPASI-10	UPASI-TRF India
2	Drought resistance	UPASI-9	UPASI-TRF India
3	Frost resistance	B-26	HPKV-TES India
4	Smallest leaf	CH-1	IHBT India
5	Biggest Leaf	Betjan	BETJAN T.E, India
6	Blister blight tolerance	TRI-2043,	TRI, Sri Lanka
7	High pubescence	TRI-2043	TRI, Sri Lanka
8	High anthocyanin pigmentation	TRI-2025	TRI Sri Lanka
9	High tolerance to pH	TN-14-3	TRF, Kenya
10	Poor fermenter	12/2	TRF, Kenya
11	Mite tolerance	7/9	TRF, Kenya
12	Scale insect tolerance	TN-14-3	TRF, Kenya
13	High polyphenol content (53.7%)	Luxi white tea	TRI, China
14	High amino acid content (6.5%)	Anji white tea	TRI, China
15	Low caffeine content (0.14%)	Guangdong tea	TRI, China
16	High caffeine content (6.96%)	Wild tea at Yunnan	TRI, China
17	Water logging tolerance	TV-9	TES, India

Improved clones		
S.No	Clone	Important features
1	UPASI 1 (Ever green)	Hardy, Quality-Above average
2	UPASI 2 (Jayaram)	Hardy, Quality-Above average, tolerant to drought and wind
3	UPASI 3 (Sundaram)	Natural triploid quality clones and very high yielding
4	UPASI 6 (Brooklands)	Suited to mid and higher elevations
5	UPASI 8 (Golconda)	Suited to all elevations, high yielding
6	UPASI 9 (Arthrey)	Fairly tolerant to drought and withstand slightly high pH, high yielding
7	UPASI 10 (Pandian)	Hardy, Quality-Above average, tolerant to drought and wind
8	UPASI 14 (Singara)	Suited to higher elevations, High yield
9	UPASI 15 (Spring field)	Flushes throughout the year
10	UPASI 17 (Swarna)	Flourishing well at mid and high elevations
11	UPASI 24	Hardy
12	UPASI 25	High yielding
13	UPASI 16	High yielding

14	UPASI 27	Drought tolerant
15	UPASI 28 (UPASI 10 * TRI2025)	Biclonal, Good strength and high quality

9.4 Climate and Soil

Tea is exacting in its climatic requirements. The temperature may vary from 16 to 32°C and annual rainfall should be 125 to 150 cm, which is well distributed over 8-9 months in a year. The minimum temperature should not go below 7°C. The atmospheric humidity should be always around 80% during most of the time. Very dry atmosphere is not congenial for tea. It is grown in plains in North Eastern States but in South India, it is grown in hill ranges from 600 to 2200 m above mean sea level.

Tea is a calcifuge crop requiring comparatively low amounts of calcium but high quantities of potassium and silicon. They can be grown in lateritic, alluvial and peaty soils. Optimum pH range is 4.5 to 5.0 and soil depth should be 1.0 to 1.5m.

9.5 Propagation

Tea can be propagated by seed and also by different vegetative propagation methods.

I. **Seed propagation**: Seed is convenient to use as planting material in tea. Till today tea is generally propagated from seed. But during recent years due to use of high yielding clonal material vegetative propagation methods are preferred. Seed collected from the fruits of seed berries are soaked in water and only heavy seeds, which sink are alone used for sowing in bed. The seed berries should not be in drought prone areas and should be at least 20m away from tea fields. Germination occurs within 20-30 days and the seedlings may be transplanted to polythene sleeves and provided with a polythene cloche. The seedlings become ready for planting in 9 months.

II. **Vegetative propagation**: Different vegetative propagation methods can be adopted in tea are:

1) By budding
2) By grafting on rootstocks
3) By layering
4) By cutting

Cutting is the most popular method of vegetative propagation as vegetative propagation by budding, grafting or by layering are too laborious to adopt.

Grafting: It is practiced for successful propagation of high yielding drought -susceptible clone used as scion on to drought-hardy clone used as root stock. In south India nursery grafts have been made using high yielding clones such as UPASI -3, UPASI - 8 and UPASI -17 as scions and drought hardy clones UPASI-2, UPASI-6, UPASI-9, UPASI-10, UPASI-24, ATK-1, SA-6 and TRI 2025 as root stocks.

(See colour version on page 210)

Vegetative propagation by cuttings: Commercial method of propagation is through single node cutting (Semi hard wood cuttings is preferred method). For raising 1,00,000 plants at a time and hardening them, an area of 1,500 m^2 land is required. Elite clone should be selected with desirable characters like yield, quality, quick recovery etc. The soil used for nursery should be free from nematodes. The cuttings callus in 4-6 weeks and rooting takes place in 10-12 weeks. The plant with good root system may be sorted out and manured with the soluble nursery mixture. They could be kept under overhead shade for 1 or 2 months and gradually exposed to direct sunlight. The plants can be fully exposed for 4-6 months when about 6 months old. The 9-12 months old plant becomes ready for planting in the field.

9.6 Main Field Preparation

Land preparation for planting tea is an important operation. Any faulty land preparation techniques may destroy the good structured soils or result in failure of the establishment of tea. For planting in virgin land, the area needs to be cleared completely by removing all vegetation and then it is prepared for planting tea. If the area was under tea or some other crops, land preparation needs extra care before replanting tea in such uprooted areas. Some techniques to be followed while planting are:

1. Level the area with minimum soil disturbance –Ploughing done if necessary. Levelling done by hand hoeing and by using leveller.
2. In case of virgin land soil testing is necessary before tea planting as tea plant does not establish properly without proper treatment.

3. In some cases, to correct the soil pH, apply lime and dolomite.
4. As uprooted lands are normally poor in fertility and soil physical properties, we have to go for soil rehabilitation by growing grasses like Guatemala before replanting.
5. As tea requires filtered shade and if it is exposed to direct sun its growth is affected, accordingly field should be chosen.

9.7 Planting

The land is cleared of the roots of the fallen trees and drains are taken at suitable intervals depending upon the slope to conserve the soil. In older days, up and down system of planting at 1.2m x1.2m are followed but presently, contour planting either in a single hedge or double hedge system is followed.

Sl. No.	Type	Spacing	Population/ha.
1.	Up and down	1.2 x 1.2m	6,800
2.	Contour planting single hedge.	1.2 x 0.75m	10,800
3.	Contour planting double hedge.	1.35 x 0.75 x 0.75m	13,200

(*Source*: www.indiaagronet.com)

The last method has many advantages over the first two, *viz*., early and high yield, better soil conservation, less weed growth in the hedge and efficient cultural practices. Planting season normally coincides with June/July and September/October for South West monsoon area and North East monsoon areas. Pits of 30×30×45cm size are dug and plants of 12-15 months old are planted by removing the polythene sleeves. Immediately after planting, plants are staked to prevent wind damage.

9.8 After Care

Immediately after planting, the soil surface around the plants should be mulched, usually cut grasses of Guatemala are employed for this purpose. About 25 tonnes of grass is required to mulch one hectare. Care must be taken to keep the mulch materials away from the collar region as they may cause collar diseases. If there is a dry weather, mud tubes may be buried 15cm deep near the plant in a slanting position and one litre of water per plant may be poured or injected at weekly intervals. This subsoil irrigation helps to minimize the causality besides encourages developing deeper roots.

9.9 Pruning

In the young tea, when it has established well, centering i.e. removing the growing point leaving 8 to 10 mature leaves from the bottom, is done to induce secondary branches. When the secondaries reach more than 60 cm, they are tipped at 50-55 cm height by removing 3 to 4 leaves and bud to induce tertiary branches. Therefore, plucking at mother leaf stage is continued for better frame development. It takes nearly18 to 20 months from planting to reach regular plucking field stage.

Pruning is done in tea

- to maintain to convenient height for plucking
- to induce more vegetative growth
- to remove dead and diseased branches

Pruning is normally done at 4 to 6 years interval depending upon the altitude of the garden, nature of the materials etc. The bushes marked for pruning should have adequate starch reserves in roots. Otherwise, the sprouting following pruning will be less. This can be normally tested by the common iodine test and if the starch reserve is less, bushes are allowed to rest for 2 to 3 months. The different types of pruning are as follows:

Sl.No.	Type of pruning	Pruning height (cm)	Season	Remarks
1.	Rejuvenation pruning	20–30	April – May	Done to old bushes affected with canker and wood rot to invigorate the new healthy branches. Not done regularly.
2.	Hard pruning	30 – 45	Apr. – May	First formative pruning done to a young tea.
3.	Medium pruning	45 – 60	Aug. – Sept.	Normal pruning where ever frames are healthy.
4.	Light pruning	60 – 65	Aug. – Sept.	Normal pruning where ever frames are healthy.
5.	Skiffing	65	Aug. – Sept.	Mainly to postpone pruning and to encourage better frame development.

(*Source*: www.indiaagronet.com)

Immediately after the rejuvenation or hard pruning, the cut ends are smeared with a paste made of copper oxychloride and linseed oil (1:1). The prunings, consisting of only small twigs and leaves are buried in trenches of 30cm width and 45cm depth taken across the slope in alternate rows. The pruned bushes are given washing with 10% lime solution using No. IV nozzle of power sprayers in order to kill the epiphytic growth of moss and lichen so as to induce early and even bud break. Lime washing also minimizes sun scorch to the bush frame.

Pruning height (cm)		Tipping height		Tipping in material	
China Hybrid	Assam Hybrid	China Hybrid	Assam Hybrid	China Hybrid	Assam Hybrid
35-45	35-55	5	4	3 leaves and a bud	4 leaves and a bud
45-55	55-60	4	3	4 leaves and a bud	4 leaves and a bud
55-75	60-75	2	2	4 leaves and a bud	4 leaves and a bud
-	> 75	-	1	-	4 leaves and a bud

The buds from the pruned shoots grow in a steady succession without any cessation of growth. These are known as a periodic shoots or primary shoots. These primary shoots should be induced to produce flush shoots, otherwise known as periodic shoots by regular tipping operation. Tipping is the removal of terminal portion of the shoot and it varies with jats and pruning height as given above. Tipping height refers to the number of leaves that must be left above the pruned cut while tipping in material refers to that portion of the terminal shoot, which must be tipped off.

9.10 Manuring

Tea responds to manuring and it has been estimated that to produce 100kg of made tea, tea plant utilises on an average 10.2, 3.2 and 5.4 kg of Nitrogen, Phosphorus and Potash per ha. Manuring in tea starts from nursery stage itself. Once they strike roots (after 4 months) 30g of soluble mixtures (Ammonium phosphate (20:20) 35 parts, potassium sulphate and Magnesium sulphate each 15 parts and zinc sulphate and Magnesium sulphate each 15 parts and zinc sulphate – 3 parts is dissolved in 10 litres of water and is applied with rosecan for about 900 plants. This must be repeated at 15 days intervals.

Nitrogen: The recommendation for mature tea is based mostly on soil organic matter status and anticipated yield. For a field with medium organic matter status the following rates of application is suggested for every 100kg of made tea anticipated:

Yield level (kg/ha)	Rate of Nitrogen (for 100 kg. of made tea)	No. of split applications
<3000	10 kg	4
3000	8 kg	5
3000 and above	9 kg	6

Potassium: Nitrogen and potassium are always applied together. NK ratio of 1:1 is used for plucking fields while for a pruned field 2:3 NK ratio is recommended. For rejuvenation pruned field 1:2 NK ratio is suggested. The enhanced rates of potassium application in the pruned year are to encourage formation of healthy plant. Muriate of potash is the sources of potassium used in tea fields. The NK fertilizers are applied by broadcast for mature tea and is broadcast and dibbled in along the drip circle for young tea. The interval between two successive applications should be at least 3-4 weeks.

Phosphorus: Phosphorus is applied once in alternate years @ 90kg P_2O_5/ha for fields yielding less than 3000kg/ha and for fields yielding between 3000 and 4500kg/ha, 60 to 80kg P_2O_5/ha is suggested every year. The soils being acidic, rock phosphate could be advantageously used. The fertilizer should be placed at 15-22cm depth.

Micronutrients: Among the micronutrients, zinc deficiency is often manifested in young shoots characterised by reduced leaf size, rosetting, chlorosis and formation of more banji shoots. Application of zinc sulphate @ 6 to 8kg/ha for high yielding fields every year is the general recommendation. The above quantity can be given in 4 to 5 split applications. It has been found beneficial to combine other micronutrients, *viz*., Manganese sulphate @ 15.5 g/10 litres and boric acid @5.5 g/10litres of spray volume along with zinc sulphate spray.

Liming: In the hill soils, due to the leaching of bases by rain and also due to the incessant application of acid forming fertilizers, the soil pH is often reduced which affects the physical and chemical properties of soil. Therefore, periodical application of lime is essential to amend the soil and maintain optimum pH. Agricultural lime (Calcium carbonate) and dolomitic lime (Calcium Magnesium carbonate) are generally recommended for tea soils. The rate of application is based on soil pH, rainfall, fertilizer usage and length of the pruning cycle. Roughly lime @ 1.5 mt/ha for a pH between 4.5 to 4.9 and 3.0mt/ha for a pH between 4.0 to 4.4 and 4.0mt/ha for a pH of less than 4.0 is suggested. The lime is applied by evenly broadcasting prior to pruning once in a pruning cycle. First manuring following liming can be had after 6 weeks and a minimum of 15cm rainfall should have been received during this period.

9.11 Shading

Conventionally, tea is grown under shade trees in India. The beneficial effects of shade trees in tea fields have been well realized. These trees help to regulate temperature and humidity at bush level. They minimize the loss of water through evaporation and transpiration. They help to reduce the injury caused to tea leaves by UV radiation. They also minimize soil erosion and increase the soil fertility by adding 8-10 tonnes of organic matter/ha/year. In North-East, trees recommended for permanent shade are *Albizia chinensis, A. odoratissima, A. lebbek, Acacia lentiartaris, Derris robusta, Dalbergia sericea*. In South India, *Grevillea robusta* is the only tree recommended for permanent shade. Initially they are planted at 6m × 6m and later thinned to 12m × 6m after 8-10 years of planting and finally into 12m × 12m by 12-15 years from planting. They are pollarded at 8-9m allowing 3-4 tiers of branches below the point of pollarding. They are subjected to annual lopping before the onset of monsoon.

9.12 Irrigation

Tea is a rainfed crop, therefore, scope for irrigation is rather limited. Measurements have shown that the average loss of water from a south facing surface is 25% higher than from the north facing surface. Subsoil irrigation may be given for young tea seedlings during summer months. Sprinkler irrigation is best for tea plantation. However, drip is suitable for areas which are newly planted or replanted.

9.13 Harvesting or Plucking

Plucking consists of harvesting 2 to 3 leaves and a bud. It is the most labour-intensive operation in a tea industry and also decides the yield and quality of made tea. Normally, a pluckable shoot takes 60 to 90 days for harvesting since its sprouting from the axillary buds. When the shoot is plucked up to mother leaf, it is known as light plucking and if it is plucked below mother leaf, it is called hard plucking.

It is essential to add one tier of active maintenance foliage to the bush every year. This is done by mother leaf plucking during January to March. During the rest of the period level plucking can be carried out. Consequent to plucking, bush height increases every year in the order of 10cm over tipping height in the first year, 7.5cm, 7.5cm, 5cm and 5cm over the previous year height in the second, third, fourth and fifth year respectively.

9.14 Yield

Yield of made tea per hectare depends upon many factors such as elevation, clonal or seedling jats, management practices, severity of pruning, processing techniques etc., Generally, in tea industry, a field which yields up to 2000kg of made tea/ha is considered as low yielding and 2000 to 3000kg as medium yielding and anything above 3000kg as high yielding fields.

9.15 Crop Protection

Sl.No	Pests	Symptoms	Control measures
1.	Tea mosquitoes (*Helopeltis antonii*)	Small adult bugs and hairy orange nymphs suck the sap from fresh leaves and tender shoots; leaves curl up, dry and die; active from January to September.	Collect and destroy bugs during the initial stages; Spray Clothianidin 50 WDG 120 g/ ha or Thiacloprid 21.7 SC 375 ml/ha or Thiamethoxam 25 WG 100g/ha or Bifenthrin 8 SC 500 ml/ha.
2.	Short-hole borer (*Xyleborus fornicatus*)	Grubs make a typical short-hole on the branches and inside galleries. A serious problem in low and mid elevation areas	1. Selectively remove badly affected branches at the time of pruning. 2. Apply nitrogen and potassium at 1:2 ratio in the prune year and mid cycle.
3.	Red spider mite (*Oligonychus coffeae*)	Infests upper surface of mature leaves	Tetradifon (8 EC) 1 to 1.25lit/ ha. The best way to control red spider mites is through the use of their natural predators like Lacewings and lady bugs but predatory mites can also be used
4.	Scarlet mite (*Brevipalpus californicus*)	Discolouration of leaves often leads to defoliation	
5.	Purple mite (*Calacarus carnatus*)	Leaves exhibit smoky grey colour	
6.	Pink mite (*Acaphylla theae*)	Young leaves turn pale and get twisted.	Dicofol or Ethion @ 1lit. /ha.
7.	Yellow mite (*Polyphagota rsonemus latus*)	Infest pluckable shoot, leaves become rough, brittle and corky in under surface.	
8.	Thrips (*Scirtothrips bispinosus*)	Leaf surface becomes uneven, curly and metty, exhibiting parallel lines of feeding marks on either side of the midrib	• Maintain optimum overhead shade. • Spray any of the following insecticides, *viz.,* malathion 50 EC 400-500 Ml, Chlorpyriphos 20 EC 500 Ml, or Profenofos 50 EC 800-1000 ml with 500 L Water/Ha.
9.	Nematodes (*Meloidogyne javanica, M. incognita)*	Occur in tea nursery, infested roots develop galls.	Pre heat treatment of soil media up to 60-80^{0}C and application of carbofuran 3G @ 80g/m^3 of medium.

Diseases			
Sl.No	Disease	Symptoms	Control measures
1.	Blister blight (*Exobasidium vexans*)	Infects tender leaves and stem and develops translucent spot. Cloudy and wet weather favours infection.	Copper oxychloride 350g in 67lit of water with power sprayer for pruned field at 3-4 days interval. In the plucking Oxychloride + 210g Nickel chloride in 45lit of water/ha at 7 days interval.
2.	Black root diseases (*Rosellinia arcuata*)	Infected roots show black mycelium on the roots, white star shaped mycelium between bark and wood and black lead shot like perithecia seen on collar region.	The soil may be drenched with Dithane M-45 @ 30g/10 litres.
3.	Red root disease (*Poria hypolateritia*)	Infected roots exhibit blood red mycelium on washing. It spreads fast but slowly kills.	1. Take trenches of 1.2m deep and 45 cm width surrounding the infected bushes and uproot and burn the bushes in situ. 2. Rehabilitate soil with Guatemala grass. 3. Soil fumigation with methyl bromide carbon-di-sulphide
4.	Brown root disease (*Fomes noxius*)	Infected root wood turns soft and spongy, it spreads slowly but kills quickly.	

9.16 Processing

Basically, there are two types of processing, *viz.*,

1. Orthodox method in which the rolling operation is done in a series of rollers. The rollers have rotary tables with battens, jacket for loading the leaf and a pressure cup,

2. CTC method (cutting, tearing and curling) which has a CTC machine, consisting of series of a pair of rollers mounted in such a way they rotate in opposite directions and the clearance between them is so adjusted to crush and tear the leaves.

Irrespective of the method, manufacturing of tea involves the following steps

Withering: The objective of withering is to reduce the moisture content of leaves by spreading them in troughs which receive artificial air from fan fitted on one end. At the end of withering, the leaves attain a flaccid condition for which it may take 12 to 18 hours depending upon the weather condition.

Rolling: This operation is carried on by a series of machines or in a single roller, during which the cells in the leaves are broken to liberate the sap containing the polyphenol oxidase, an enzyme, which in the presence of oxygen, oxidises the polyphenols to produce theaflavins and thearubigens. These are responsible for colouring of the tea and it takes 30-40 minutes. Afterwards, the fine sifted rolled ones are sent for fermentation while the coarse ones are again sent for rolling.

Fermentation: Rolled tea materials are either spread in concrete floors or kept in aluminium trays. In the presence of high humidity proper step decides the quality, i.e., strength, colour and briskness of tea. Fermentation requires 1 hour or 2 hours depending upon the environmental conditions.

Drying: This step aims at stopping the fermentation process and slowly removing the moisture content without a burnt smell but preserving the inherent quality. This is achieved by passing the fermented tea in thin layers through conveyors into a drier in which the inlet temperature is maintained around 250 – 280^{0}F and outlet temperature is a round 150-200^{0}F. Proper drying takes 30-40 minutes.

Grading: Before grading, the dried tea is removed of the stalky fibres, which affect the quality, by passing through fibre separate machines. The bulk tea is passed through different sized meshes which aid in separation into different grades.

Orthodox grades	Mesh size	CTC Grades	Mesh size
Pekoe	>8 mesh sieve	Flowery Pekoe (FP)	>8 mesh
Tippy golden orange pekoe (TGOP)	8-12	Pekoe	8-10
Broken orange pekoe (BOP)	12-16	BOP	10-12
BOP – Fannings	16-18	Pekoe Fannings	12-16
BOP –dust	18-24	BOP – Fannings	16-20
Dust – I	25-30	Pekoe dust (PD)	20-30
Dust – II	Below 30	Red Dust (RD)	30-40
		Super Red dust (SRD)	40-50
		Finel dust (PD)	50-60
		Superfine dust (SD)	Below 60

References

ICAR 2010. *Handbook of Horticulture.* ICAR, New Delhi.

Sharangi AB 2009. Medicinal and therapeutic potentialities of tea (*Camellia sinensis* L.) – A review. *Food Research International*, **42**: 529-535.

e-resources:

https://www.semanticscholar.org

http://www.agritech.tnau.ac.in

http://teaworld.kkhsou.in

http://indiaagronet.com

http://www.printfriendly.com

http://www.statista.com

www.ibef.org

OUTCOME ASSESSMENTS

PART A

Answer the following questions true or false.

1. Tea is the native of China in South East Asia. True/False
2. Basically, there are three types of processing in tea. True/False
3. Tea can be propagated by seed only. True/False
4. Tea is a rainfed crop. True/False
5. Tea is an evergreen plant. True/False

PART B

Answer the following questions.

1. Conventionally, tea is grown under — trees in India.
2. Define tipping height.
3. Why pruning is done in tea?
4. How tea is propagated?
5. To correct the soil pH — and — are applied.

PART C

Write a brief note on each of the following.

1. Processing of tea.
2. Manuring in tea.

3. Pruning of tea.
4. Harvesting of tea.
5. Climatic and soil requirements of tea.

10

Production Technology of Date Palm

Botanical Name: *Phoenix dactylifera* L.
Family: Arecaceae
Chromosome No: 2n = 36
Place of Origin: Countries around Persian Gulf

10.1 Introduction

Date palm is a nutritive fruit rich in sugar and iron and predominantly seen in desert oasis. It is the main fruit crop in arid and semiarid regions, particularly in the arid regions of western Asia and North Africa. The palm tree is well adapted to desert environments that are characterized by extreme temperatures and water shortage. Beyond the arid climates, date palm can also be grown in many other countries for food or as an ornamental plant including the continents of Americas, Southern Europe, Asia, Africa and Australia. The majority of date palm growing areas are located in developing or underdeveloped countries where dates are considered the primary food crop, thus playing a major role in the nutritional status of these communities. By-products from date palm are used in building structures, animal feed, and also in several items such as baskets and ropes. The date palm tree has been in cultivation since 2400 BC. It is one of the oldest cultivated fruit trees on earth. India is the leading importer of dates. According to FAO statistics, in 2011 India was the largest world importer of dates (256,295 mt) and its neighbour Pakistan, in 2012, was the world's sixth largest producer (600,000 mt). The major date palm growing areas in India are Kutch (Gujarat), Rajasthan and certain parts of the Punjab, as well as Tamil Nadu state to some extent. In Kutch, there are more than 2 million date palms, the majority of them grown from seeds and offshoots, providing a huge biodiversity for experimentation and improvement of products. Efforts by the Indian government in the last decade have resulted in an increase of date palm cultivation from 8,973 to 16,000 ha and establishment of a prominent effort in Rajasthan (Shah, 2014).

The date fruits contain moisture ranging from 10 to 22%, total sugars 62 to 75%, protein 2.2 to 2.7%, fibre 5 to 8%, fat 0.4 to 0.7%, ash 3.5 to 4.2%, total acidity 0.06 to 0.20%, and ascorbic acid 30.0 to 50.0 mg per 100 g edible portion on dry weight basis (Baraem *et al.* 2006; El-Sharnouby *et al.* 2007). The flesh of dates contains 0.2–0.5% oil compared to 7.7–9.7% in the seed. Unsaturated fatty acids of the fruit include palmitoleic, oleic, linoleic and linolenic acids and the oleic acid content of the seeds varies from 41.1 to 58.8%, indicating that the seeds of date could be used as a source of oleic acid. They contain at least six vitamins including a small amount of vitamin C and vitamins B_1 (thiamine), B_2 (riboflavin), nicotinic acid (niacin) and vitamin A. Dates contain 0.5–3.9% pectin, which may have important health benefits (Al-Shahib and Marshall, 2003).

They contain antioxidant flavonoids such as ß-carotene, lutein, and zea-xanthin. These antioxidants found to have the ability to protect cells and other structures in the body from harmful effects of oxygen-free radicals. Thus, eating dates is found to offer some protection from colon, prostate, breast, endometrial, lung, and pancreatic cancers. Zea-xanthin is an important dietary carotenoid that selectively absorbed into the retinal macula lutea, where it is thought to provide antioxidant and protective light-filtering functions. It thus offers protection against age-related macular degeneration, especially in elderly populations.

Beside direct consumption of the whole dates the fruits are traditionally used to prepare a wide range of different products such as date juice concentrates (spread, syrup and liquid sugar), fermented date products (wine, alcohol, vinegar, organic acids) and date pastes for different uses (e.g. bakery and confectionary). Again, the by-products arising from date processing can be used for different purposes. Within agricultural systems date press cake (by-product of date juice production) as well as date pits can be used as animal feed-stuff (also dates falling down from palms before maturity). In case no other uses are possible, all organic waste material arising from date processing shall be used as a component for compost preparation at least.

10.2 Origin and Distribution

It is believed to have originated in countries around Persian Gulf such as Iraq (Mesopotamia) and Egypt. Date palm has been cultivated in the Middle East and North Africa (MENA region) for millennia. The African date palm (*P. reclinata*) or the Indian date palm (*P. sylvestris*) or both may have been the progenitor of date palm. Date palm can be cultivated in all five continents of the world, largely between 39° northernmost and 20° southernmost latitudes. However, the main region of production is the Middle East and North Africa, where 89% of dates are produced (Zaid and de Wet 2002).

10.3 Choice of Cultivars

Date includes 200 genera; the genus *Phoenix* contains 12 of the 1500 species that belong to the date palm family. *Phoenix* palms are dioecious and are characterized by pinnate leaves and in duplicate leaflets with acute tips. Besides date palm, the other two most highly valued *Phoenix* palms are Canary Island Palm (*P. canariensis* Chabeaud) an ornamental palm and the Sugar Palm (*P. sylvestris* Roxb) that is common in the Indian subcontinent for its sugar syrup. Date palm can be distinguished from the other two species by the production of offshoots and the dull glaucous leaves instead of glossy leaves.

Depending on the season of ripening, the cultivars are classified as early, mid and late cultivars. Though there are nearly 40 cultivars imported from Middle East and North Africa, few only were found to be promising, under Indian conditions.

Hallawi: It is soft date from Iraq. An early cultivar with small fruits. At full maturity before ripening (doka stage) the fruits are yellow in colour and free from astringency. This cultivar is relished both in doka and dang stages. Total soluble solids range between 28 to 42% and astringency in the fruit at doka stage is low or almost absent. It is an early cultivar yielding good crops and is somewhat more tolerant to rains. Dry dates (Chhuhara) of good quality and cured soft dates (Khajoor) can be prepared. Its fruit at dang stage (fresh form) is very delicious. Average yield ranges from 50 to 80 kg per palm.

Khadrawi: It also originated from Iraq. This cultivar proved successful under Abohar conditions. The trees of this cultivar are comparatively less tall and yield good crops. Small to medium sized fruits; light yellow at doka stage. It can be used both for dry dates and soft dates. However, due to high astringency at doka stage, it is not suitable for eating as fresh. TSS at doka stage varied from 29.5-42 % at Abohar, 34% at Jodhpur and 36.5% at Bikaner. The yield ranges from 40 to 70 kg per palm. This cultivar required heat summation unit of 2231^{0}C and 3342 ^{0}C for reaching doka and pind stage during first fortnight of June and second fortnight of July, respectively at Bikaner.

Shamran: It is a mid-season cultivar tolerant to high humidity. It is soft to semi dry date cultivar and highly astringent at doka stage. The cultivar yields prolific crops. The fruits can be cured into good quality Chhuhara and also softened with salt treatment. Fruit is medium to large, oblong to oblong-oval and yellow at doka stage. It recorded very low berry weight (6.01-12.89g) under Indian condition. Its yield per palm is 40 to 70 kg.

Medjool: A late cultivar originated in Morocco and has large size, broad, oblong-ovate, orange yellow with reddish brown fruits stippling at doka stage.

It is particularly good for preparing dry dates of attractive bold size and good quality but not suitable for raw eating. The fruit is large sized having much variation in weight. The mean berry weight recorded at Jodhpur was 16.2 g at Bikaner and 19.57 g at Jodhpur. The TSS at doka stage varies from 31% to 36.9% depending on climate and region. This cultivar required 2648 ^{0}C for reaching doka stage at Bikaner. Its yield is 35 to 50 kg per palm.

Barhee: It is soft date from Iraq. Mid-season to slightly late cultivar with small to medium sized fruits, yellow colour and low astringency at doka stage. The cultivar has proved *extremely good for table purpose in fresh form (doka stage*). The doka fruit contains about 32% TSS, has golden yellow colour and has a very pleasant taste because of low astringency and high pulp content. The trees of this cultivar are prolific in yields and the fruit is ready for eating late in the season i.e. first fortnight of August at Abohar. Its yield ranges from 60 to 110 kg per palm.

Hayany: It is soft date originated from the UAE. The berries have attractive deep colour. The fruits could be consumed as fresh in the doka stage as ripening does not proceed further under Abohar conditions. The yield per palm ranges from 30 - 40 kg.

Zahidi: It is a mid-season cultivar, slightly tolerant to rain or high humidity. The fruit is small to medium, obovate and yellow at doka stage. It is a semi dry date cultivar from Iraq not suitable for raw eating, though most suitable for processing to prepare pind khzoor. Mean berry weight recorded 10.4 g. It required 2322 ^{0}C and 3479 ^{0}C to reach colour turning stage and pind stage respectively at Bikaner.

Khalas: It is a mid-season cultivar. Small to medium size oblong oval, yellow fruit and sweet at doka stage. It has an oblique base and irregular outline. It is suitable for both fresh consumption and for processing as soft dates.

Deglet Nour: Unique taste and mostly known in the Middle Eastern. Semi-soft and famous flavour, light to dark brown in colour, harvested semi-dry.

10.4 Climate and Soil

Dates can be successfully cultivated in areas having long hot summer and mild temperature during flowering (February to April) and fruit ripening (May to August) should be 25 ^{0}C to 29 ^{0}C. As in the case of grapes, this crop also requires specific heat units (above a base of 10 ^{0}C nearly 3000 units) for successful fruit maturation. There should not be any rains or high humidity during fruit maturity and ripening, otherwise the fruits will be spoiled and cracked. The crop is susceptible to frost.

Deep sandy loam is the best suited though it is very hardy and can be grown in a wide range of soil conditions. The date palm produces its best crops only with proper drainage and soil aeration. Its roots descend into loamy soils to a depth of 2-3 m; therefore, proper soil aeration to this depth is to be maintained. Even if irrigation is present, the permanent water table level should be at 2.7-4.9 m. It can grow in alkaline and saline soil; however, the growth and productivity are affected.

10.5 Propagation

Though it is propagated by seeds, the seedlings exhibit a very long juvenile phase nearly taking 7 - 8 years for flowering. Besides this, the population segregates for male and female plants, resulting all female plants also may not be uniform in bearing. Hence vegetative propagation through off-shoots produced by a female mother plant of particular cultivar can be used for planting. After about 4- 6 years, 2-3 off shoots can be obtained from a mother palm for a period of 8-10 years. Each matured off-shoot may weigh 25-35 kg.

For offshoot propagation, if the fruits are ripe on the mother tree, at that time the offshoot is cut, there can be no doubt of the type of fruits that the offshoots will bear. Before being removed, offshoots should be pruned of mature leaves, preserving the tender young opened leaves in the central bud. To prevent excessive water loss from transpiration, offshoots should weigh at least 2.7 kg; those of lesser weight will likely die. Ideal weight of an offshoot for planting is 15-20 kg and be 3 to 4-year-old. To encourage rooting, the base of off-shoots arising from the mother plant are applied with moist soil by putting a soil boxes at least for a period of one year before separation. The ideal time of off-shoot removal from mother plant is spring (Feb-March) and autumn (August-September). The off-shoots with sufficient roots are planted in the field or nursery. When planting the off-shoot, a cleared circular area 5-6 m in diameter around each tree is recommended. Adjacent rows should alternate and not be opposite. Development of tissue-cultured date palms has led to further expansion of date orchards in many parts of the world.

10.6 Planting

Pits of 1 × 1 × 1 m sizes are dug during summer and filled with a mixture of 20 kg FYM, 2 kg castor cake, sawdust, sand and methyl parathion 2% dust. Planting of off-shoots is done during rainy season. Most of the basal leaves are removed and only the terminal buds with 10-12 leaves around them are retained. Depending on the soil fertility, the spacing varies from 4-9 m. Commercially, spacing of 6-8 m is adopted. In India a spacing of 3-4 m is adopted. For effecting pollination under commercial cultivation 2-3 male plants are planted for every 100 female plants.

10.7 Water Management

Light and frequent irrigations have to be given immediately after planting and the basins can be mulched. Offshoots in well drained medium loamy soil in early September will require daily watering for the first 40 days, every-other-day watering for the next 40 days and thereafter watering every 6 days until growth begins. Adult palms need to be irrigated every 15 days during winter and every 7 days in summer. Though date palm is drought tolerant, 2-3 m around root zone is to be kept moist for maximum growth. The irrigation frequency varies with season. Date palm is highly tolerant to saline water (even up to 2500 ppm). But to overcome drainage problem arising out of saline water usage, periodical leaching with food water should be done.

10.8 Nutrition

An annual application of 23 kg of well-seasoned manure, 200 g each N, P and K per adult plant is needed. The whole quantity of FYM, P, K and half dose of N is applied during the month of September-October and remaining half in January-February. Green manuring with any quick growing and easily rotting leguminous cover crops is always beneficial. Green manuring crops are generally grown during monsoon. In Punjab, coal ashes appear to be the best for spreading around the plant as they form good mulch and prevent termite attack (Shah, 2014).

10.9 Intercrop

Intercropping can be practiced during the early years; if the field has irrigation facility, farmers can get some extra income by having inter-crops in between the dates palm plants. Inter-crops like green gram, black gram, peas, lentils, vegetable crops, papaya and pomegranate can be cultivated. In case of inter-cropping, make sure to have additional requirement of water and fertilizers.

10.10 Training and Pruning

To get optimum yield 75-100 leaves are necessary. Generally pruning is not required up to the initial 5 years as about 20 leaves are produced each year. However, older leaves which are in surplus than required are normally pruned. The leaf pruning is done during June. By this pruning, the bunches will be better ventilated during July and early August, which will otherwise develop black nose spots on the fruit. The spines from the leaves around the bunches are also cut during spring to facilitate pollination. One cluster should be retained to every 7-8 leaves in Zahidi, Khadrawy, Barhee and 8-9 leaves in Halaway, Deglet Noor and Dayri. Optimum yield and good quality fruits are obtained with 8 active leaves per bunch when 5-8 bunches/plant are to be retained.

10.11 Flowering and Fruit Set

Date palm is a dioecious plant; that is, staminate (male) and pistillate (female) flowers are borne on two different palms. Male and female flowers are arranged in strands that attach to a rachis forming an inflorescence called spadix. A bract, called spathe, enclosing the immature inflorescence, splits longitudinally at anthesis, which allows for the pollination of mature male and female flowers.

Pollination: For better yield hand pollination is done. The timing of hand pollination varies; if the winter is warmer, flowering is earlier. Pollination is appropriate when the female spathe bursts and the stigma is receptive. The male flower cluster with its enclosing spathe is cut when it is about to open. Waxy scales cover the stamens. One or two small branches are cut from the male cluster and placed among the small branches of the female cluster. One male tree may suffice to pollinate 100 female trees; however, it is safer to have 2-3 male palms per 100 females.

Though the dried pollen can be stored at 4-5 ^{0}C until next season, the fresh pollen produces the best fruit. The spathes emerge during February-March and flower opening starts during March - April. Immediately, the flowers should be pollinated (2-3 days after spathes open). Pollen grains of certain cultivars can advance ripening of certain cultivars. So specific polliniser cultivars should be identified for specific cultivars of female plantation.

Fruits: Fruits of date palm, called dates, develop from one fertilized ovule forming one carpel, while the other two ovules are aborted but remain visible at the fruit calyx. If no fertilization occurs, three or more carpels develop simultaneously. The date fruit develops on the flowering strands and is a berry characterized by a membranous endocarp surrounding a seed. Large variations exist in the shape, size, colour and chemical composition of date fruit depending largely on varietal differences but also on climate, soil and growing conditions (Al-Yahyai and Al-Kharusi 2012).

Following pollination and fertilization, date palm fruit grows through five distinct developmental stages characterized by physical and chemical changes at each stage. These stages are Hababouk, Kimri (Ganodara), Khalal (Doka), Rutab (Dang) and Tamar (Pind) (Zaid and de Wet 2002). The Hababouk stage lasts 4-5 weeks after fruit set and is characterized by slow growth rate. The fruit is round and the colour is cream to light green. The Kimri stage lasts 9-14 weeks and during this stage the fruit elongates and gains rapid development and increased weight, volume and reducing sugars. The fruit at the Kimri stage is green and is marked by high tannin concentration that is reduced as the fruit develops. The Khalal stage, also known as Bisir, is when the fruit is physiologically mature and the colour changes to the cultivar specific colour, usually various shades of red and yellow. This stage lasts 3-5 weeks and is marked by slow growth (3%-4% weight gain) to fruit full size and weight, and further accumulation of sugars. Several cultivars with high sugars can be consumed either fresh or boiled at this stage. The Rutab stage is when the fruit apex starts to soften and the colour changes to a darker colour. Most dates are consumed fresh at this stage that lasts 2-4 weeks. Due to softening, dates at this stage lose water content to an average of 30-45%, but sugars continue to accumulate. The Tamar stage is when the date palm fruit is fully ripe and has completely softened. The fruit colour darkens at this stage, which is also marked by low moisture content (10-25%) and high concentration of total sugars. This stage lasts 2-4 weeks and the dates are appropriate for long-term dry storage or processing.

Fruit Thinning: To improve fruit quality and to ensure their proper nutrition and development on the bunch, thinning is necessary. It has to be resorted to so as to retain 1300 - 1600 fruits in 8-10 bunches per palm and it will be optimum. Bunch thinning can be done either by removal of entire strand or shortening of strands. Spraying ethephon @ 200 ppm 10 - 30 days after fruit set will help to thin fruits effectively. It also helps to overcome biennial bearing and encourage earlier ripening and to get better fruit weight and soluble sugar.

There are three methods of fruit thinning: (i) cutting the strands to reduce their length; (ii) removing some strands from the centre of the bunch or (iii) reducing the number of fruits on each strand. Experiments showed that removal of one-third of the strand from the centre in all bunches improved fruit qualities and ripening in cv. Barhee.

Seedlessness: Spraying of GA_3 @ 25-100 ppm just before pollination or at time of flowering of female flowers may induce high percentage of seedlessness. The seedless fruits are similar in fruit shape, size, colour and most of their physical and chemical characteristics.

Fruit ripening management: Spraying of ethephon @ 200 ppm leads to early ripening but decreases fruit size, whereas application of GA_3 @ 400 ppm delays ripening. However, combine application of both ethephon and GA_3 results in formation of large size fruit and early ripening in March.

10.12 Crop Protection

10.12.1 Diseases

1. *Alternaria* leaf spot (*Alternaria alternate*): This disease was recorded from moderate to severe form depending upon cultivars as well as climatic conditions in Western Rajasthan and Kachchh area of Gujarat. Disease incidence was noticed in KVK, Fatehpur (Rajasthan) up to 45-50%.

Symptoms: The gray to brown powdery spots are most common on the lower leaves (pinnae) of the plants, whereas, on upper leaves the spots are few and small in size. The disease causes heavy losses to the date industries in both quality and quantity of production. The fungus perpetuates as mycelium in infected crop debris. The infection starts from the fruits on which numerous conidia are formed. These conidia are spread by wind, rain splash and cause secondary disease. The conidia germinate in the presence of moisture at temperature nearly 25-30 °C giving rise to germ tubes, which enter the host tissue. It requires hot and humid weather conditions for disease initiation. Disease development is favoured between 20 to 30 °C. High relative humidity coupled with rainy days favours the disease development.

Management: Application of sulphur-based fungicide is best to control the disease.

2. Fusarium wilt (*Fusarium oxysporum*): *Fusarium* wilt in date palm was first reported from Qassim region, Saudi Arabia in 1990. This disease has been observed in date palm growing areas.

Symptoms: The fungus attacks old date palms as well as offshoots and young palms. Spines begin to whiten from bottom to top followed by brown discoloration along the midrib of the leaves. Symptoms extend from the outer leaves to the young ones, then to the apical meristem, leading to the death of the date palm tree. A discoloration of the parenchyma and vascular tissues occurs in the transverse section of affected leaves. Pathogen grows in the vascular system and blocks the flow of nutrients and water. Leaves dry due to fungal infection. The pathogen can survive in crop debris in the field. Soil temperature and moisture are known to affect the disease development. Disease is severe at temperature between 23-30 °C. Light intensity and higher RH coupled with high evaporation rate increase the disease development.

Management: An effective therapeutic method for Fusarium wilt is not available. Therefore, prevention is the best approach to combat this disease. Nursery and entry point inspections prevent or at least slow down the introduction of the disease from other states. Make sure pruning tools are disinfected before moving into another palm. Some success has been achieved in selecting resistant cultivars from amongst the natural palm population and in breeding trials. The low level of genetic variation found in the pathogen suggests that improved, rapid and specific detection methods based on molecular techniques may be relatively easy to develop for quarantine purposes and that resistance breeding could be accurately targeted.

3. Fruit rot (*Aspergillus niger*): Disease incidence was recorded during survey program from 10.0 and 30.5% in Bajju and Bikaner (Rajasthan), respectively.

Symptoms: The rot begins as small, circular spots which enlarge and turn brown to black spores. Fruit rot damage varies from one year to another depending on the humidity and rain and time from the Khalal stage until fruit maturation. Phialides directly develop on the vesicle which forms by swelling of conidiophores tip. Conidia are formed in chains. These conidia cause infection by air and rain splashes. Rotting is more when humidity is high and temperature is between 25-30 °C.

Management: Lowering the humidity inside the bunch, by the use of wire rings, and/or by removing a few fruit strands from the centre of the bunch, will facilitate ventilation and drying of wet fruit. Protection from rain or dew is reached by using paper covers in the early Khalal stage to cover the fruit bunch. Fungus spoilage could also be limited by dusting the fruit bunches during the Khalal stage with 5 % ferbam, 5 % malathion, 50 % sulphur and an inert carrier (40 %).

4. Diplodia disease: Diplodia disease, caused by *Diplodia phoenicum* (Sacc), has been recorded on 20 date cultivars all around the world, although it appears to be most common to Deglet Nour.

Symptoms: Disease is severe on offshoots and is characterised by death either while they are still attached to the mother palm or after they have been detached and planted out. The fungus may infect the outside leaves and finally kill younger leaves and the terminal bud or the central cluster may be infected and die before the older leaves. Yellowish-brown streaks extend along the leaf base.

On the leaves of older palms, the ventral mid-portion of the stalks is commonly affected, showing yellowish brown streaks, 15 cm to over one meter in length, extending along the leaf base and rachis. The upper part of the leaves however, may still appear green and unaffected.

Management: Since the fungus usually enters the palm through wounds made during pruning or cutting when removing the offshoots, one precaution is to disinfect all tools and cut surfaces. Dipping or spraying the offshoots with various chemicals (benomyl, Bordeaux mixture, methyl thiophanate, thiram and other copper-based fungicides), has been found effective against the disease.

5. False smut or graphiola leaf spot: The causal organism is *Graphiola pheonicus*.

Symptoms: Small, dark brown to black spots appear on affected leaves. These spots are scattered on both upper and lower surface of leaflets. Under high humid condition this disease is intensified.

Management: Remove all affected leaves and destroy them. Application of Bordeaux mixture at a ratio of 4:4:50 and its repetition after 15 days interval is advisable.

6. Inflorescence rot (and/or Bayoud): Triggered by *Fusarium oxysporum* f. sp. *Albedinis*. This fungus exists in the soil.

Symptoms: White chlorite colour and fade of the palm leaves. Bad conditions of cultivation and an intensive cultivation of alfalfa and vegetables in rotation supports an infection. The following cultivars have a lower fruit quality but they are supposed to be resistant against inflorescence rot:

Takerboucht, Bou Jigou, Taadmant and Bou Stammi.

10.12.2 Insect-Pests

1. White scale (*Parlatoria blanchardi*)**:** It is widely distributed throughout most of the date growing regions of the world. It is generally believed to be native to the Arabian Gulf countries. Commerce in date offshoots over the centuries paved the way to spread of this pest into India, Central Asia and Africa. Date palms are the preferred host of scale, but it has also been recorded on other hosts belonging to four plant families: Arecaceae (Palmae), Apocynaceae, Oleaceae and Rhamnaceae.

Damage: It mainly infests the leaves and sucks the sap. If full leaf is covered with white scales, finally the leaves dry out. In heavy outbreaks, fruits are also attacked and fall off before maturity. Nymphs and adults suck the sap from the leaflet, midribs and the fruits. Under each scale insect, a discoloured area appears on the leaflet. Heavy infestation causes leaflets to turn yellow and contributes to premature death of the fronds. Respiration and photosynthetic activities are almost stopped resulting in early death of the infested leaf. Damage on fruits is easily noticeable and the production is not marketable, especially due to poor acceptability. The infestation of date palms scale, *Parlatoria blanchardii* begins in December, and peaks in October. The infestation begins on the basal tissues of the pinnae and moves upwards. Older leaves and upper leaf surfaces near the pinane are preferred, and the tips remain free of infestation. Infestation of the pinnae declines in May-June and is concentrated on the floral parts and berries. Crawlers are most active in February-April.

Management: The best protection can be achieved by covering date cluster with alkathene. This covering also protects from birds and damage caused due to rain. Natural enemies and pruning normally keep pest populations at tolerable levels.

2. Caterpillars / Lesser date moth: Most common are the caterpillars of the butterfly *Ephestiaca utella* and *Batrachedra amydraula.*

Damage: Larvae are found to damage the large number of young fruits from April to June or July. They eat the leaves and penetrate into the fruits. The larvae attack newly formed inflorescences, but the main injury is to the young green fruits. Approximately 80% of the fruits are attacked when between 0.6 and 1.0 cm in diameter. The larva chews a hole near the calyx, through which it penetrates the fruit and feeds on the pulp and the soft immature seed. A damaged fruit is easily recognizable by the black faeces attached to the penetration site. The larva also moves from fruit to fruit within the bunch, thus increasing the damage. A larva seldom eats more than one-third of a fruit before seeking another one, and damage about three to four fruits during its life span. About 4 weeks of infestation, the fruits become dark, dry and fall off. In severe infestation most of the infested fruits drop to the ground; the bunch ceases to grow, and then dries. Thus, lesser date moth injuries to the palms cause considerable fruit drop and losses of up to 75% of the yield. The first larvae appear in April to start the damage on the newly formed fruits.

Management: Bunch remnant pruning and bunch covering, when applied along with chemicals, greatly decrease pest infestation in Iran. Date palm cultivars (like "Sammany") with a relatively higher wax content are more tolerant to the pest than those with less wax (like Hayany). Furthermore, the tolerant cultivar has a better structure of the outer and inner layers of fruits tissues.

This moth has developed much resistance to pesticides, including organophosphates. Plastic strips impregnated with the organochloride dichlorvos (20%), suspended within the bunch, also provide satisfactory control, as do insect growth regulators or spraying of chloropyriphos @ 2ml/ lit of water controls the larva population.

Biological control: Two larval ectoparasitoids attack *B. amydraula* in Israel, *Bracon* sp. (Braconidae) and *Parasierola swirskiana* Argaman (Bethylidae). The former attacks larvae of various lepidopteran species; its effect on pest numbers is not known. *Parasierola swirskiana* occurs only in the south of Israel, attacking the pest's second-generation larvae in the early spring. Inundative releases of the egg parasitoid *Trichogramma cacoeciae* Marchal (Trichogrammatidae) are continuing. Anthocoridae, mostly *Orius* spp., as well as spiders, contribute to the natural control of *B. amydraula* in immature dates. Various products of *Bacillus thuringiensis* Berliner (Bt), were effective in controlling the pest, especially those infesting small fruits of which the pests consume several. An Israeli-developed Bt blend, "Bitayon", based on 20% concentrated liquid mixed with wheat flour and yeast extract (as a phagostimulant), added to date pollen, provided date fertilization as well as protection for small fruits. The blend is harmless to parasitoids and predators.

3. Red palm weevil /Coconut beetle: The red palm weevil, *Rhynchophorus ferrugineus* Oliv also called the Indian palm weevil, is well known in the Middle East where it causes severe damage on date palms. Approximately, 5 to 6 % of palms in the Middle East region are infested with the red palm weevil with an annual rate of infection of about 1.9%.

Damage: Infestation is often not apparent until extensive damage has already been caused and the palms are beyond recovery. These infested plantations show wilting symptom with yellow inner leaves and rotting odour could be smelt from the damaged tissues. Small round holes at the sites of removed offshoots are also a clear indication of the presence of the weevil. Chewed up date palm fibres are extruded and a brown fluid oozes out of the holes on the stem. Cocoon, weevil and pupal fibres are frequently found in the palm leaf base. Heavily infested palms may contain 80 or more simultaneously developing larvae, and more than one generation may develop in the same palm. Infestation may adversely affect the potential yield, lower the palm's growth rate and eventually cause its collapse and death. After emergence, the adult weevil stays inside the cocoon for approximately 8 days, until attaining sexual maturity.

Management: This can be controlled by destroying any decaying logs in palm plantation by chopping and burning. The larva can be killed by filling kerosene oil or any insecticide into the holes and plugging the holes. Removing broken tree or stump is necessary. It is advisable to remove all infected trees.

4. Rhinoceros beetle (*Oryctes rhinoceros* L.): It is one of the serious and important pests of date palm in India.

Damage: When the attack is on the unopened spathe, the inflorescence becomes badly damaged and can cause 10 % annual reduction in yield. Frequent infestation results in reduction of leaves and stunting of trees. The damage can cause death in seedlings and young palms but adult palms can withstand infestations. The adult beetle burrows and remains between leaf sheaths near the crown and cuts the leaf in the folded stage thereby causing permanent damage. The major injury to the palm is caused by the grubs, which feed on the roots and bore into the underground bases and even the trunks. Severe damage is inflicted particularly to offshoots and young palms, in which the mortality rates may be very high. Mature and old palms are also infested with grubs, which results in yellowing of the palms and reduction in yield.

Management: Same as red palm weevil.

5. Termites *(Microcerotermes diversus)*

Damage: Termites usually feed on cellulose matter and the attack starts from the root zone and base of the offshoots by making vertical canals through it, or building soil-canals on it, allowing them to reach the stem. Where termites are found, they usually cause the death of newly planted offshoots. They may also make galleries in the trunks of weak palms and cause them to collapse.

Management: Control measures could be started by removing and burning destroyed offshoots. In case of a slight attack, it is recommended to clean the offshoot of soil canals and spray it with a termite killer. It is also advised to turn over the surrounding soil to about 50 cm deep in order to destroy these canals and treat them with a nematicide product (which will certainly kill all termite species). Termites in young plantation can be controlled by application of BHC 10% dust or spraying of chlorpyriphos @ 2ml/lit of water after cleaning of soil canals.

6. Bugs: Rhinoceros bug of the species *Oryctes* spp. They eat tissue of the young leaves and destroy the area of vegetation.

Management: Removal of their hotbeds like rotten plant material and green manure. Artificial preparation of hotbeds for catching, biological control by the fungus *Metarrhizium anisopliae* and the virus *Rhabdionvirus oryctes*.

7. Rodents: As in other palm crops rats, mice and other rodents may cause damages on the trunk as well as on the fruits. Feeding on offshoot roots affects their survival. They also feed on roots of old palms causing them to fall down if feeding is only on one side of the palm and wind is severe.

Management: It is recommended to support predators like owls with the objective to control the population of rodents in the date plantation. Another mechanical way of reducing fruit damages is to place a mechanical device around the stem in order to make it impossible for rodents to climb up the tree. Rat can be controlled by using poison, a mixture of zinc phosphate at 30 to 50 g with 1kg of millet flour and 3 % of cooking oil. The paste is to be placed around the palms at the entry to the galleries.

10.13 Harvesting and Post-Harvest Management

Under Indian condition, since the fruit ripening period is not free of rains, the fruits have to be harvested at doka stage during June-August. With proper care, the date palm will produce fruits from the fourth year onwards and yield economically satisfying amounts after 7-8 years, continuing as an adult tree of 60-70 years. Annually, an average adult palm will yield 100 kg of khalal stage fruits and an elite cv. up to 200-300 kg.

Table 10.1: Different stages of development of fruit

Name	Arabic name	Stage	Fruit quality
Ganodara	Kimiri	4-13 Weeks after pollination	Hard, green colour.
Doka	Khalal	13-17 weeks after pollination	Hard, yellow, pink or red, may be sweet or astringent, edible stage.
Dang	Rutab	17-21 weeks after pollination	Fruits soften at tip, edible stage.
Pind	Tamer	21-22 weeks after pollination	Fully ripe, 60-84% TSS, edible stage.

Date fruits are eaten at *doka* or khalal stage, soft or ripened stage (*pind* or tamar) and at dry stage (*chhuhara*). India produces and markets dates at the khalal stage because climatic conditions do not favour full ripening on the tree to produce tamar dates. Netting to protect the ripening fruits is necessary to protect against bird damage.

Sardarkrushinagar Dantiwada Agricultural University (SDAU, 2010) made the following observations about date growing in Gujarat:

(a) Khalal stage of Halawy cv. is the best quality for making *chhuhara*, prepared by immersing fruits in boiling water for 20 min and subsequent drying on trays in an air circulation oven at 45° C for 60-65 hr.

(b) Barhee cv. is the best for making *pind khajoor*.

(c) Medium quality fruits are useful for preparing good quality beverages. The Local Red cultivar is good for making jam.

(d) Cultivars Medjool, Hatemi, Ruzaiz, Selection-3 and Tayar are resistant to white scale, *Parlatoria blanchardi*.

(e) Use of 600-gauge low density polythene at colour break stage of the fruit (ICBR of 1.1.73) protects against rain damage in late maturing cultivars.

Date curing: As doka and dang stage fruits cannot be stored for future use, curing should be done. For chhuhara preparation from doka stage, fruits are boiled for 5-10 minutes depending on cultivars and then dehydrated in solar dryer or oven at 48-50 °C for 70-95 hr. Fruits harvested at doka stage can be artificially ripened to bring them final stage of maturity by dipping in boiling water for 20-25 seconds and then dehydrated in oven at 38-40 °C.

Fruits can be stored for a longer duration in cold storage. Ripe fruits can be stored for 2 weeks at 5.5-7.5 °C with 85-90% relative humidity. Whereas, dried fruits can be stored for one year at 1-11 °C with 65-70% relative humidity.

References

Al-Shahib W and Marshall RJ 2003. The fruit of the date palm: its possible use as the best food for the future? *Int. J. Food Sci. Nutr.,* **24**: 247-259.

Al-Yahyai R and Al-Kharusi L 2012. Physical and chemical quality characteristics of freeze-stored dates. *International Journal of Agriculture and Biology*, **14** (1): 97-100.

Baraem I, Imad H, Riad B, Yehia M and Jeya H 2006. Physicochemical characteristics and total quality of five date varieties grown in the United Arab Emirates. *Int. J. Food Sci. Technol.,* 41: 919-926.

El-Sharnouby GA, Al-wesali MS and Al-Shathri AA 2007. Effect of some drying methods on quality of palm Date fruits powder. The Fourth Symposium on Date Palm in Saudi Arabia. King Faisal University, May 5-8.

Shah JJ 2014. Date palm cultivation in India: An overview of activities. *Emir. J. Food Agric.*, **26** (11): 987-999.

Zaid A and de Wet PF 2002. Botanical and systematic description of the date palm. In: A Zaid and EJ Arias-Jiménez (eds.), *Date Palm Cultivation. FAO Plant Production and Protection Paper 156 Rev. 1.* Rome, Italy.

OUTCOME ASSESSMENTS

PART A

Answer the following questions true or false.

1. Date palm is a monoecious plant. True/False
2. Spraying of GA_3 just before pollination may induce high percentage of seedlessness. True/False
3. The flesh of dates contains 0.9–3.5% oil. True/False
4. Date palm is a nutritive fruit rich in sugar and iron. True/False
5. An average adult palm yields 300 kg of khalal stage fruits. True/False

PART B

Answer the following questions.

1. What is *doka* stage?
2. In date palm generally pruning is not required up to the initial — years.
3. Under Indian condition the fruits have to be harvested at *doka* stage, why?
4. How can you control *Alternaria* leaf spot?
5. Why hand pollination is done in date palm?

PART C

Write a brief note on each of the following.

1. Propagation of date palm.
2. Water and nutrient management in date palm.
3. Flowering and fruit set in date palm.

11

Production Technology of Rubber

Botanical Name: *Hevea brasiliensis*
Family: Euphorbiaceae
Chromosome No: 2n =36
Place of Origin: Amazon basin in Brazil
Common Name: Rubber tree, Rubber plant, Para rubber tree

11.1 Introduction

Rubber is a fast-growing tropical tree and mainly grown for the production of latex. In 1495, Columbus was the first person to report about latex. The latex was used for rub out pencil marks and this gave the product name as rubber. The latex is used for production of various rubber products. The most important application of latex rubber was initially accomplished in the 1880s when the rubber was used for making motor cars tyres. When rubber is heated with sulphur, it retains the physical properties.

Among different species, *Hevea brasiliensis* is superior to all other types and was first reported by H. Ridley. The genus *Hevea* is native to South America and found as wild species in the Amazon and Orinoco regions. Around 99% of the World's natural rubber production comes from *Hevea brasiliensis*, the rest natural rubber is extracted from the shrub guayule (*Parthenum argentatum*). *Hevea* lies between latitudes 25° N (Yunnan Highlands, China) to 21° S in Brazil, but the commercial worldwide production lies between 15° N and S latitudes. *Hevea brasiliensis was* introduced to tropical Asia in 1876. The first rubber plantations in Malaysia were established as early as 1890. Now, the crop is commercially grown in Asia, Africa and America, Indonesia, India, Sri Lanka, China etc. The tropical Asia plays major role in rubber production. Today, most latex production is concentrated in industrial estates in tropical Africa and the far East. In India, the rubber cultivation was started in the year 1905 in Periyar and Poonoor districts of Kerala. More than 90% of the crop area is cultivated in the states of Kerala and Tamil Nadu. The cultivation now extended to non-traditional areas like Karnataka, Assam, Meghalaya, Maharashtra, Goa and Odisha. The rubber production in India is not sufficient and it is able to meet only 40% of its requirements. The rest 60% is imported from other countries. Indian industries mainly depend on natural rubber whereas international market is mainly dominated by synthetic rubber.

11.2 Botany and Classification

Rubber belongs to the family of Euphorbiaceae, a large family with about 280 genera and 8,000 spp. The genus *Hevea* exhibits much morphological variability, with nine species till now recognized. All the species contain latex in their parts in different concentrations. Other *Hevea* spp. are in wild state and of little economic importance. Some of these species can be used for breeding programme for crop improvement. The characteristic features of some species are:

Sl. No.	Species	Native	Characteristics
1	*H. benthamiana*	Northern Brazil, Colombia and Venezuela	A high-quality latex which is slightly lower in yield than *Hevea brasiliensis*
2	*H. camporum*	Amazon basin	Shrub
3	*H. guianensis*	Rainforests of Ecuador, Venezuela, the Guyanas, Brazil, Colombia and Peru	A yellowish latex of inferior quality to *Hevea brasiliensis*
4	*H. microphylla*	Amazon basin where it occurs in Venezuela, Colombia and northern Brazil.	The poor quality of the latex which is mixed with resins
5	*H. nitida*	Rainforests of northern Brazil and Colombia.	Thin latex that has the undesirable quality of preventing coagulation (anticoagulant) when mixed with the latex from other rubber tree species
6	*H. pauciflora*	Rainforests of Venezuela, the Guyanas, northern Brazil, Colombia and Peru.	Latex is low in quality and being mixed with high resins
7	*H. rigidifolia*	Rainforests of northern Brazil and Colombia	Cream coloured latex having a poor reputation due to the high proportion of resins present
8	*H. spruceana*	Rainforests of northern Brazil and Guyana	Watery latex is almost devoid of rubber

Hevea brasiliensis is a quick-growing tree and it grows up to height, 25 m. The plant has well-developed strong tap-root system and 30 to 60% of feeder roots are concentrated in the top 10-30 cm of the soil. The feeder roots absorb all the nutrients and water from soil and later transported to other parts. The trunk is conical or cylindrical in shape. The trunk consists of a central cylinder of wood and is surrounded by bark. The bark is the significant portion of the tree as it produces the latex. It consists of pith, wood and cortex. The cortex is further divided into 3 separate concentric layers, i.e., the outer corky layer (periderm), an underlying parenchyma and the phloem with the latex vessels. The wood tissue is covered with bark and it produces latex on tapping. The xylem tissues are found inside and phloem tissues on the outsides. The mature bark remained in the inner zone and these are soft in nature. The intermediate zone is hard in nature whereas the outermost consist of a layer of cork cells. The soft tissue consists of laticifers which contains latex. Though the hard bark contains laticifers but they are discontinuous and non-functional. The leaves are green in colour, trifoliate in nature with long petiole. The young leaves are dark red in colour. The nectar is secreted in flush of leaves during flowering. The leaflets are elliptic or obovate in shape. In the beginning of the dry season, the older leaves shed and the tree becomes bare for a short period which is known as "wintering". Trees older than 3-4 years show annual shedding. The inflorescence is borne in the axils of new shoots, many branched panicles consisting of flowers of both sexes. Flowers are small, scented, unisexual and shortly-stalked. Petals are absent. The duration of flowering is of about two weeks. Male flowers are smaller than females and male flowers are in the ratio of 60-80 males to each female flower. The females are white, sticky and sessile stigma. The fruit set in rubber is less and out of these 30-50% fall off after a month. Seeds are viable for a short time only. The fruit takes 5-6 months to mature after fertilization. The mature fruit is a large, compressed, 3-lobed capsule, 5cm in diameter, with 3 oil-containing seeds. Viable seeds germinate in 3-25 days. Germination is hypogeal. The pollination in *Hevea* is entomophilous. Bees, midges and thrips help in pollination.

11.3 Soil

The rubber plant is hardy in nature and it can be grown in diverse range of soils. The plant has well developed root system. So, it requires deep, fertile, well-drained soil for proper development of root system. The soil should have good water holding capacity and water table should be at a depth of 100 cm. The most favourable pH for growing rubber lies between 4.0 to 6.5 but the crop can be grown in the pH range of 3.8 to 8.0. The young seedlings are more sensitive to low pH level. The growth of plant is retarded in case the pH exceeds 8.0. In India, the crop can successfully be grown in lateritic, alluvial and clayey loam soils.

11.4 Climate

As a tropical crop, it can naturally grow between 15° N and 10° S latitude. The crops can be properly grown in wet equatorial type of climate. It is a low altitude crop and grows adequately up to a height of 600 m above MSL. Leaf diseases and low yield have been recorded if the crop is grown above 300 m MSL. It is the difficult to grow rubber in slopy land as there is erosion risk, difficulty in movement of tappers and maintenance of estate is not easier. The ideal climatic conditions for rubber comprise of several ideal conditions, i.e., stable mean temperatures of 20-28^0 C, a well distributed high annual rainfall of 1800-2000 mm, a relative air humidity between 60-80% and 1,500 to 1,800 hours bright sunshine per year. Low temperatures lead to stunted growth and the latex production decreases in the long run. When rainfall goes over 2,500 mm, fungal infections are seen in the leaf and tapping is disrupted as the heavy rainfall dilutes the latex, causing it to fall on the soil. Dry spell for 2-3 months may be helpful and can encourage extensive re-foliation. The perfect rainfall pattern is that, when the rain fall starts in the late afternoon and stops just before tapping the next morning. *Hevea* is moderately susceptible to wind. Strong winds can uproot trees or break stems. No rubber plantations exist therefore in cyclone-prone areas.

11.5 Choice of Clones

The clones are classified based on the method adopted for the development of their mother trees. These clones are broadly classified into three categories, i.e.,

Classification	Clones
A. Primary clone: Selection of mother tree is based on superior performance. The parentage is unknown.	Tjir-1, GT1, PB 86, PR 86, PR107 and PB 28/59
B. Secondary clone: The mother tree is evolved by controlled pollination between 2 primary clones.	RM 600, RR II 105, RR II 208, RRIM 628, PB 5/51
C. Tertiary clone: Mother tree is evolved by controlled pollination where at least one or both parents are secondary clones.	RRIM 703, RRIM 707, RRIM 501, RRIM 623

Rubber Research Institute of India and other institutes have developed clonal varieties. Some of the important clones are:

RRII 105 CLASS I	The clone was developed by the Rubber Research Institute of India and its parentage is Tjir I and GI 1. The stem grows straight with dense canopy at the top. Leaves are glossy. Late self-pruning occurs but in exposed areas light branches may appear drooping. The latex is white and the dry rubber production is around 2425 kg/ha/yr. The clone shows slightly tolerance to abnormal leaf fall but highly susceptible to pink diseases.

RRIM 600 CLASS I	The clone was developed by the Rubber Research Institute of Malaysia and its parentage is Tjir I and PB-86. Stem straight with moderate branching, leaves are obovate, light green and leaflets are separated. Initially the plant grows slowly with delayed branching. These branches appear alternately. Canopy is dense with light green colour. The dry rubber production is around 2,102.54 kg/ha/yr. This clone shows tolerance to pink disease.
GT 1 CLASS I	It is a primary clone grown extensively in Indonesia and other countries. Trunk upright and erratic branching pattern. Latex is white with annual yield of 1400 kg/ha/yr. The clone shows slightly tolerance to pink diseases and mild abnormal leaf fall.
PB 217 (Rb 99-04) CLASS II	The parents of Malaysian clone are PB 5/51 and PB 6/9. The trunk is tall and straight. The leaves are V-shaped, obovate and light green. In the initial stage the stem looks dark brown in colour which grows erect with smooth leaf-scars on the surface. Latex is pale yellow. Crown is conical fan-shaped and wind resistant. The latex is yellow with dry rubber production of around 2,161.17 kg/ha/yr. The incidence of powdery mildew disease is severe.
RRIM 712 CLASS II	The leaves are oval, boat-shaped and dark green in colour. Stem is dark brown with presence of prominent leaf-scars on the surface. Flat branching with crown shaped canopy. The Latex is yellow with dry rubber production of around 1,645.40 kg/ha/yr.
RRIM 901 CLASS II	The leaves are obovate and narrowing towards the base with dark green wavy leaf margin. Dense with healthy conical canopy, branches are well-arranged and the lower branches are drooping in nature.
PB 235 (Rb 99-02) CLASS II	The leaves are elliptical boat-shaped with sharp tips and colour is dark green. Petioles are thick and light-coloured leaflets are touching each other. The tree grows vigorously to form a conical crown with semi-erect branches. The trunk grows straight that makes good quality timber. Latex is pale yellow with dry rubber production of around 2,241.74 kg/ha/yr.
PB 260	The leaves are oval and boat shaped, dark green colour with wavy leaf margin. Petioles are thick and well clustered. Erect and smooth trunk, producing good timber. Thin, virgin and renewable bark. Conical crown is high-set. Early self-pruning occurs. Some have acute secondary leader. Latex is pale yellow with dry rubber production of around 2, 370.0 kg/ha/yr.
USM 1 (Rb 99-03)	The leaves are oval-shaped, leaves with separated leaflets, light green coloured leaves. Vigorous tree with erect and smooth trunk with dry rubber production of around 2,498.15 kg/ha/yr.
PB 311 CLASS II	Leaves are light green, convex, obovate and well separated. Smooth trunk with fan-shaped crown. The dry rubber production of around 1,590.74 kg/ha/yr.
RRIM 628	The parentage of this clone is Tjir 1 and RRIM 527. Leaves are glossy, light green, oval-shaped and with well-arranged petioles. Initially it grows slowly. The tree produces a light green dense canopy. Dry rubber production of around 1,504.60 kg/ha/yr.
RRIM 703	The parents of this clone are RRIM 600 and RRIM 500. The latex colour is light yellow. The average annual yield is 1310 kg/ha. This clone is severely prone to abnormal leaf fall. The occurrence of powdery mildew is mild but the clone is prone to severe pink disease.

11.6 Propagation

The rubber is propagated through seed and by bud grafting. Now-a-days it can be multiplied by micropropagation or tissue culture technique.

11.6.1 Seed Propagation

The viability of seeds remains for 3-4 days if exposed to sunlight but under shade condition it can be retained for 7 days. The viability of seeds can be maintained up to one month by keeping the seeds in well aerated containers along with wet charcoal having 40% moisture. The viability of seeds can extend up to 4 months by storing the seeds at 4^0C in polythene bags. Seventy percent viability can be retained up to one month by packing the seeds. Plantations raised from seeds are having low potential yield. Fruits normally ripen in the month of July to September. The fruits are harvested from the plant when they turn to yellow-brown stage and seeds are collected by breaking them. Seeds are sown in the raised nursery beds with 1.0-1.2 m width and convenient length. Nurseries need to be established in flat lands and near to a water source as young plants need to be irrigated daily. The nursery bed should be prepared taking soil, sand and farm yard manure in equal proportion. Around 6-7 days after sowing the seed germinates under favourable condition.

Nurseries are prepared for raising seedling, budded plants and bud wood. Nursery bed site should be in the open space and the land must be levelled. Beds should be well drained fertile soil, free from weeds and watering is done as per soil moisture status. Organic mulching helps better germination, weed control and maintains the soil moisture. In nursery the poorly developed seedling can be removed and homogeneous population can be maintained. For a plantation of 100 ha about 1ha of nursery is needed. The germinated seeds are planted in the nursery at 30 × 30 cm, giving approximately 100,000 plants per ha. When the seedling attains a height of 3-4 feet in six months, the entire plant is dug or pulled out of the ground, tied into bundles, transported to the field and planted in the main field in monsoon season.

Hevea plants are budded with improved clones in a nursery. Prior to transplantation in the field, the rootstocks are cut obliquely above the budding point to stimulate the growth of the clone.

11.6.2 Vegetative Propagation

It is an alternative propagation method by budding. The seedlings raised from seeds show great variation in the latex yield. Buds are generally taken from high yielding trees and grafted on seedling in order to get the bud graft plant. Buds are required for bud grafting and success of budding can be ascertained after 3-4 weeks after budding. Brown budding is done by taking the healthy buds from one-year bud wood plant and the budding is carried out in the stock plants that are of 10-12 months old having at least 7.5 cm growth at the base. The bud wood plant of one-year growth produces 20-40 buds when the plants attain a height of 1-2 m. Rainy season is the best time for budding. The desirable buds are collected from the axils of the fallen leaves. In case of green budding, the bud and the stock plant are young. Green buds are taken from the axils of leaves of bud stock plants that are of 6-8 weeks old. Vigorous seedling of 2-8 months old having 2.5 cm girth at base are used as root stock. As per requirement the bud wood nurseries are established and are of two types, i.e., brown bud wood nursery and green bud wood nursery. The place that is selected for bud wood nursery is first cleaned properly and leveling is done as per requirement. In case of brown bud wood nursery, the spacing is 90 × 60 cm whereas in green bud wood nursery the spacing of 1 × 1 m is maintained. Time to time fertilization and other activities such as irrigation, mulching, weeding, shading, protection against diseases and pests are carried out in order to get proper growth of the plant. Care should be taken during dry weather. The age of stock plant and bud wood should be of similar age. In case of field budding two or more seeds are planted at stake and the vigorous seedling is budded. If the first budding fails, budding is carried out in the second seedling.

11.6.3 Polybag Nursery

Polybags are used for growing germinated seeds and the plant is bud-grafted when it attains five to six months. Budded stumps can also be raised in the polybags and the budded stumps are treated with indole butyric acid (IBA) to enhance proper root growth. The black polyethylene bags are generally preferred to transparent bags. Low density polyethylene (LDPE) sheet of 400 gauge is used for making small bags and 500-gauge thickness polyethylene is usually used for making large bags. The high-density polyethylene (HDPE) sheets can also be used for preparation of bags but it deteriorates when it is exposed to light. Polybags of larger size (65 cm × 35 cm) and small size (55 cm × 25 cm) are used. The larger bags hold about 23 kg soil whereas small bags hold about 8 to 10 kg soil. Larger bags are used for producing six to seven whorls. Sufficient number of holes is made in the lower half of the bags to facilitate the drainage. The soil having good structure, clay-loam texture and with good water holding capacity are ideal. It is better to keep the filled bags in trenches. Trenches are made in such a way that they should be equal to the diameter of the bag. In certain cases, these trenches are dug in such a way that two pairs of bags can accommodate at a time. In between the trenches, 75 cm gap was maintained for foot path. The soil mixture bags are placed in the trenches and then the gap space needs to be filled up with soil. The sprouted seeds and budded plant are planted in the poly bags. In case of budded plant, the buds should face towards the road side for proper growth of the sprouts. During the first month of establishment, NPK Mg 10-10-4-1.5 mixture is applied @10 g per bag and gradually the fertilizer dose is increased up to 30 g in four months' time. Irrigation should be done regularly as per requirement. Special care should be taken to avoid water logging. Partial shade may be given to the plants by constructing overhead shade.

11.6.4 Nursery Management

Nursery should be grown with utmost care for production of quality planting material. The nursery area should be free from weeds and as per the requirement, 3-4 hand weeding is carried out in a year. Pre-emergence weedicide can be applied in order to avoid the first hand weeding. Application of diuron @ 2.5 kg/ha in 700 L water is done after preparation of the nursery bed to control the weeds in the initial stage. Then, after 5 days of spraying the germinated seeds are sown in the beds in July to September. Mulching is done with dry leaves, grass cuttings and dry cover crop materials before the summer season. Black polythene sheets can also be used for mulching. The nurseries have to be irrigated at regular interval for proper growth of the nursery plants and the water requirement varies with soil, climate and age of plants. In the initial stage daily watering is done and the frequency of irrigation may be reduced to 2-3 times in a week. The seedlings are allowed to grow and later on budding is done or it is used as seedling stumps.

11.7 Land Preparation and Planting

Rubber plantations are generally grown in slopy area for latex production and for soil conservation purpose. The rubber planting depends upon the types of land that are used for rubber plantation. If the rubber plantations are grown in undulated areas, soil conservation measures have to be taken. In hilly areas, planting is done along the contour lines i.e., across the slope. The square system of planting is adopted in flat lands. Pits of size 75 cm^3 to 90 cm^3 are prepared depending upon the soil and the top soil and sub-soil should be kept separately, the pit is exposed to sunlight for few days and then the pit filling should be done with top soil as far as possible. Pits should be filled up with 10-20 kg organic matter and 150 g of rock phosphate for proper initial establishment. In low lying areas drainage channels are made to remove the excess water from the plantation.

Planting season starts in June and July. Around 420-445 plants /ha are maintained in case of budded plants and 445-520 plants/ha in seedling plants are planted in the main field. The planting can be done in 4 types.

A. **Seed at stake planting**: In situ planting of 2-3 germinated seeds may be planted per pit in a triangle. The plants may be thinned out later and most vigorous plants are allowed to grow. The field budding is carried out at appropriate stage in the vigorous plant.

B. **Stump planting:** The plant is planted in the field as soon as it was pulled out from the nursery. While planting care should be taken so that the bud patch should remain above the ground level in order to avoid the infection.

C. **Poly bag planting:** In this method, the planting hole is made bigger than the size of the poly bag. Bottom of the bag is removed entirely. Then the bag is placed in the hole. The plant is planted in the hole without disturbing the earth ball. After planting the soil is properly packed around the root zone. In order to reduce the adverse effect of the sunlight, the bud faces towards the North –East direction.

D. **Budded stump planting:** Seedling raised in nurseries are budded and transplanted after pruning the stem at about 8 cm from the bud patch.

At regular interval the field inspection needs to be carried out and the false shoot are removed from time to time. Only the vigorous one should be allowed to grow. The side shoots that are developed up to 2.5 m needs to be removed.

11.8 Nutrition

Manures and fertilizers help the plants to attain proper growth and early establishment. The seedling nursery is basal dressed with 25 kg of well rotten FYM and 3.5 kg rock phosphate per 100 m^2 of the nursery bed. Later these nursery beds are top dressed with 5.5 kg urea per 100 sq. m. After 6-8 weeks of nursery planting, the nursery bed is fertilized with @25 kg per 100 sq. m N-P-K-Mg mixture during September-October. Immature plants are fertilized in circular band with about 30 cm around the base of the plant leaving at least 7 cm from the base of the plant. In initial stage of the growth the plants are fertilized with 100 kg N-P-K-Mg (11-10-4-1.5) per ha consisting 440-450 plants/ha. Later stage 200-250 kg of the fertilizer is applied at regular intervals, i.e., 9 months, 15 months, 21 months, 27 months, 33 months and 39 months. The fertilizer dose is applied once in September-October and other in April-May. Fifth year onwards, fertilizer recommendations for rubber is 40 g N ,90 g $P_2 0_5$ and 90 g K_2O per plant per year. For matured rubber trees under tapping apply NPK 10:10:10 grade mixtures at the rate of 900 g/tree (300 kg/ha) every year in two split doses. Add 10 kg commercial magnesium sulphate for every 100 kg of the above mixture if there is magnesium deficiency. Fertilizer needs to be applied in square or rectangular patches in between rows. Care should be given that each patch is serving four trees. Elevated level of N and Mg can negatively influence on the quality of the concentrate latex. Too much of Cu and Mn also shows a harmful effect on the oxidative process of rubber. Higher concentration of Mg and Ca in the plant can cause unsteadiness in the latex vessels and thus dropping the time of flow and yield.

11.9 Mulching, Shading and White Washing

In order to increase the water and nutrient holding capacity of the soil, the plant basin is covered with dry leaves, cover crop cuttings, grass cuttings, paddy straw etc. It is a recommended practice. Mulching in rubber plantations is also done to protect soil, increase the soil moisture content and control the weed population. Generally, November is the ideal time for mulching to protect the plants from adverse effect of drought. Mulching can be undertaken in nurseries and young plantations after fertilizer application and before the onset of regular summer. Mulching improves the soil microbial population, facilitates better nutrient availability to the plant.

White washing is done in the main stem of young plants to protect it from the scorching heat. This process is continued till the canopy of the plants develops and thereby the stem portion can be protected through partial shades. However, plants on the roadsides may need white washing for a longer period as they are more exposed to sunlight. Lime or china clay is used for white washing the plants.

11.10 Irrigation

Rubber performs best under adequate moisture condition. The latex is water based and therefore adequate moisture supply is needed for good vegetative growth and latex production in rubber. Rubber thrives well, if the rainfall is evenly spread round the year and 1750 mm per annum is considered ideal. Good growth depends on the availability of adequate moisture and nutrients to the seedling. Seedling should be mulched during the late rains or just before the commencement of dry season.

11.11 Intercropping

During the initial period of a rubber plantation, the land area is not fully occupied by the rubber plants and inter-spaces are available that receive plenty of sunlight. These interspaces can be utilized for growing different intercrops to generate additional revenue in the initial stage of establishment. Intercropping is only possible where the land is levelled or having gentle slope. As an intercrop, banana can be grown with a spacing of 2 m × 2 m accommodating 1200 plants per ha. In the initial period banana plants are planted in single row and only one sucker is allowed to grow in the second year. Other crops like pineapple, ginger, turmeric can be taken up during initial years of establishment with proper nutrient management. Intercrops should be planted at least 1.5 m away from plant bases.

11.12 Cover Crops

In the rubber plantations with moderate to steep slopes, leguminous crops are grown as covers crops in order to check the soil erosion and for improvement in the soil status. Cover crop reduces the runoff loss and enhances the soil moisture content. Generally, cover crops like *Pueraria phaseoloides, Mucuna bracteta, Calopogonium mucunoides, Centrosema pubescens* can be taken up. Land should be prepared before onset of monsoon. Therefore, treatment is done to ensure proper growth of the plant. Acid treatments with concentrated sulphuric acid is done for 10-20 minutes for crops like *Pueraria phaseoloides, Mucuna bracteta, Calapogonium mucunoides*. Hot water treatment for a period of 4-6 hours at 60-80° C for seeds like *Calopogonium mucanoides, Puegraria phasoloides* enhances the germination percentages.

11.13 Weed Management

Weeds compete with the rubber plants for moisture, nutrients, light and thus affect the growth plants. In a well-established rubber plantation plant canopy coverage creates heavy shade thereby greatly reduces the weed growth. It is important that weed competition should be minimal during initial stages. Cover crops such as *Pueraria phaseoloides, Mucuna bracteta, Calopogonium mucunoides, Centrosema pubescens* can reduce the weed population to a great extent. Weed can either be controlled by hand weeding or by application of pre-emergence and post-emergence weedicide. Contact herbicides like Diuron @ 2 kg/ha, Simazine @ 3 kg/ha can be used as pre- emergence and translocated herbicides like Glyphosate @ 2 litre in 400 ml water as post-emergence weedicide to control the weeds, and the soil-acting herbicides (ex. Atrazine) are mainly used in nurseries to control the weeds.

11.14 Nutritional Disorder

In rubber plantations major deficiency symptoms are seen due to lack of nutrients like magnesium, potassium, zinc and manganese. In case of magnesium deficiency there is development of chlorosis (yellowing) in mature leaves and per-coagulation of latex on the tapping panel occurs due to excess magnesium content. Marginal chlorosis seen in the older leaves and reduction in leaf size are due to deficiency of potassium. Internal chlorosis of leaves, young leaflets become incurved due to deficiency of zinc. The overall paleness, yellowing of leaf with bands of green tissue outlining the midrib and main veins are due to manganese deficiency.

11.15 Crop Protection

11.15.1 Insect Pest Management

Scale insects (*Saissetia nigra* Nietu)

Scale insects are small, oval with an outer black covering distributed on rubber leaflets, petioles and terminal shoots are normally attacked by these coccids. They suck the juice from these plants' parts. Severely affected portions gradually wither and finally dry up. Ants are associated in the insect affected colonies to feed on the sugary honey dew secreted by scale insects. Later on, sooty mould develops on the affected parts and the affected portions gradually turn black. Spraying of malathion or dimethoate 0.05% will control the pest.

Mealy bug (*Ferrisiana virgata* Ckic)

These soft bodied insects suck the sap of the tender shoots and terminal leaves. The affected portions look whitish due to the presence of mealy powder. The sugary substances secreted by the mealy bugs attract ants /black ants by symbiotic process. The ants/ black ants protect the mealy bugs from their natural enemies and in turn feed on the honey dew that oozes out from the mealy bugs. In case of severe infestation, the terminal leaves as well as tender shoots dry up and give a blackish appearance due to the development of sooty mould fungus. Trimming of the infested branches in the initial stage reduces infestation. Systemic insecticide like dimethoate can check the infestation.

Termites (White ants) (*Odontotermes obesus* Rambar)

Termites are soil-inhabited social insects and remain by constructing underground colonies therein. Normally they are saprophytic in nature and feed on the dead tissues of plants as well as roots. Sometimes the young plants are damaged due to the activities of termites in the vicinity of the root zone. Keeping the rubber plantation clean and providing frequent irrigations decrease the termite population. Destruction of the termite nests and drenching it with chlorpyriphos (0.1%) particularly in the summer months decreases termite attack. Treating the soil with neem products and spraying Azadiractin to the termite pest affected rubber trees at 20-30 days interval keep the plantation free from termite attack.

Cockchapfter grub (*Holotrichia serrala* F., *Hrufaflava brenske* and *Anomala varians* Ol.)

These are soil inhabiting insects and the grub usually causes damage to the root zone in the seedling plantations. Due to root injuries the seedlings wilt and dry up. Deep summer ploughing before planting the seedling also exposes the grub to its natural enemies. Soil treatment with phorate 10 G @ 25 kg/ha and irrigating the field kill the grubs that remain underneath the soil. Adult chafter beetles are attracted towards light. Installing light traps in the rubber plantations is beneficial in checking the pest.

Bark feeding caterpillar (*Aetherastics circulate* Merr.)

The grubs after emergence bore into the trunk though weak spots. The boring tunnels provide a congenial atmosphere for the grub to feed and grow. In severe infestation the affected plants exhibit sticky appearance and the growth is affected. Locating the entry holes and injecting CS_2, kerosene, petrol or chloroform into the infected tunnel and plugging the entry point with mud or cement kills the grub inside the plant. Injecting 0.02% chlorpyriphos or quinalphos 0.05% is effective against the grubs. The grub can also be destroyed by injecting a blunt GI wire and pushing it randomly in the tunnel.

11.15.2 Disease Management

Abnormal Leaf Fall Disease

This is caused by *Phytophthora palmivora,* a universal fungal pathogen. On petioles, infections are commonly found and it is the entry point. Leaves normally fall from the petioles at this stage by turning to reddish-brown. Brownish, circular water-soaked lesions are seen on the leaves. Lesions finally combine to form big, asymmetrical necrotic areas and leaflets are easily dropped on forceful shaking. Contaminated leaves fall off in huge numbers prematurely. Spraying of 1% Bordeaux mixture (300 to 400 l/ha) and for aerial spraying 8 kg of dispersible copper oxychloride powder 56% in 40 litres oil per ha is used to control the disease.

Brown Rot Disease (BRD)

Brown rot disease (BRD) is one of the vital diseases that ruins the root parts of the rubber tree. This disease is caused by a basidiomycete fungus, *Phellinus noxius*. The symptoms of the brown rot disease are slowing the plant growth, yellowing and wilting of leaves, branch dieback, and plant death. The dead wood is started to discolour, reddish-brown and finally becomes crumbly, dry, and white. A cluster of white mycelia can be found between the sapwood and bark.

Powdery Mildew

The infections are caused by *Oidium heveae* and it causes discoloration, defoliation of young shoots, curling the margins on older leaves. The powdery mass of fungus covers the whole upper and lower surfaces or appear in patches only on the leaves. Only the petioles are attached to the twigs giving a broom-stick appearance. After a few days, the petioles will slowly fall followed by the dieback of twigs. In the older leaves, these white patches may produce necrotic spots and decrease the photosynthetic efficiency. Rubber flowers are more susceptible to *Oidium hevea* than the leaves, which results in a poor fruit pod set. Alternate use of Bavistin 0.05% a.i., (1g /l litre of water) and sulphur (2g/lt of water) at fortnightly interval controls the disease.

Shoot Rot

The disease is caused by the fungus, *Phytophthora palmivora* (Butl) *and P. meadii* Mc Rae. The disease attacks to nursery seedling plants and young plants in the field. The infection becomes severe during south west monsoon period. Repeated spraying with Bordeaux mixture or oil-based copper fungicides at regular interval controls the disease.

Leaf Spot

The disease is caused by the fungus *Corynespora cassiicola* and a severe threat to *H. brasiliensis*. The symptoms are generally appeared to be circular, rarely irregular amphigynous lesions on the leaf lamina. A spot of white or light brown papery at the centre with a dark brown ring at the margin of the leaf. Young leaves show the shivering at the leaf tips. Upper leaves are defoliated and drying of the terminal portion is common during the dry weather. Even though mature leaves are affected but very small size lesions are formed. Severe infection causes dieback on the shoot tips. In bud woods nurseries, leaf spots are common and defoliation is rarely seen. Spraying with Mancozeb (0.2%) or Bavistin (0.2%) is recommended for the nursery plants.

Pink Disease

The disease is caused by the fungus, *Corticium salmonicolor*. The plants that are of age group of 3-12 are more susceptible. In the fork region the white or pink mycelial growth are seen on bark surface. The latex oozes out of the lesions and later the infected bark dries up. Sometimes cracking is seen in the affected bark. Application of Bordeaux paste up to a height of 30 cm above and below the affected region controls the disease.

Bird's Eye Spot

It is a severe disease caused by *Drechslera heaveae*. Small necrotic spots with dark brown margin and pale centre may be observed in the infected plants. Severe infection leads to premature defoliation and later die-back may be seen. The disease may be controlled by repeated spraying of Bordeaux mixture 1% or Dithane M-45 0.2% or Bavistin 0.2%.

Patch canker or bark canker

The disease infects the tapping panel region or the collar region. It causes swelling of roots and amber-coloured liquid oozes out on brushing of bark. It is better to scrap the affected portion and the wound portion is thoroughly washed with Dithane M-45 0.75% (10 g / litre of water).

11.16 Harvesting

Latex is found in the latex vessels in the form of small particles and that flows out when tapping is done on the bark portion. Tapping is a controlled wounding process in which thin shaving of the bark is removed. Generally, the rubber plant takes 7 years to attain proper girth. In tapping the latex vessels are cut out open so that latex flows easily. In the initial stage the coagulum needs to be removed for better flow of latex. A plantation is ready for tapping when 70% of the population attains the standard girth of 50 cm for seedling plants and 55cm in case of budded plants. The best month for tapping of new plantation is March. The tapping is done early morning (2 to 6 hours) in order to get maximum latex. Tapping at late hours reduces the exudation of latex. The tapping cut also called the tapping panel and should be at a slope of 30^0 in budded plant and 25^0 in seedling tree. The yield of latex is maximum when tapping is done at a depth of less than 1 mm close to the cambium in such a way that the cambium should not be injured. Panel is the area of bark in which tapping cut is situated and helps to explain the tapping. Panel notation represents the panel position and renewal succession of the panel. Common panel notations are: BO1(A): First basal panel of virgin bark, BO2(B): Second basal panel of virgin bark, BI-1(C): First renewed bark of BO-1, BI-2(D): First renewed bark of BO-2, BII-1(E): Second renewed bark of BO1 and BII-2(F): Second renewed bark of BO2. At three days interval tapping is generally done. A skilled person can tap 300-400 trees per day. The latex is collected by using the coconut shell or by polythene cups. At 2-3 hours interval, the collected latex from the cup is transferred to a clean bucket. *Hevea* latex vessels in the latex vessels of tapped trees contain 30-40% rubber in the form of particles. Hevein is a protein that coagulate the latex at the cut- ends of latex vessels and it leads to cessation of latex flow. The binding property increases in acidic pH. The implements that are required for tapping are Michie Golledge knife, spouts, cup hangers, collection cups and collection buckets. Ethephon (2, chloro ethylphosphonic acid) is used as a yield stimulant. The Ethephon undergoes rapid degradation to phosphonic acid, ethylene and chloride ions and therefore, it is a harmless yield stimulant. Initially the latex is slightly alkaline or neutral but instantaneously it becomes acidic due to bacterial action. The fresh latex cannot be kept as such without coagulation. The pre coagulation of latex can be prevented by application anticoagulant like ammonia, sodium bisulphate and formalin. The latex that is collected from the cup is generally processed to crepe rubber, sheet rubber, processed latex, latex concentrate or block rubber. Preservatives such as ammonia (1%) or a mixture of boric acids (0.2%) and lauric acid (0.3%) and ammonia (0.2%) are used to preserve field latex. The bark regeneration mostly depends on genetic character of the variety, soil and climatic factors and also the intensity of tapping.

11.17 Tapping Systems

The reaction to diverse tapping systems depends upon high yielding clone or low yielding clone. It is also dependent on high yielding seedling plant. Generally, half spiral alternate daily (S/2 d2) system adopted in case budded trees, and half spiral third daily (S/2 d3) system is generally done in seedlings plants. In case of high yielding clones and medium yielding clones, low frequency tapping systems can be adopted. The standard system of tapping S/2 d2 and the intensity is considered as 100%. Some tapping symbols and their meaning are as follows:

Sl No	Symbols	Means
1	S	One full spiral cut
2	V	One full V-cut
3	S/2	One-half spiral cut
4	S/4	One-fourth spiral cut
5	2×S/4	Two one-fourth spiral cuts on the same tree
6	S/2 U	One-half spiral cut tapped upwards
7	2×S/2 DU	Two half spiral cuts, one tapped upward and the other tapped downward
8	d1	Daily tapping
9	d2	Alternate daily tapping
10	d3	Third daily
11	d6	Weekly tapping
12	d2 d/7	Alternate daily, six days in tapping followed by one day rest

11.17.1 Low Frequency Tapping (LFT)

In order to achieve sustainable yield and to reduce the cost of production, the frequency of tapping is reduced. When the frequency of tapping decreases that leads to reduction in the cost of wages of the tapper, which plays a significant role in cost of production. Trees under low frequency tapping (d3, d4 & d6) have to be stimulated from opening onwards for achieving maximum. Tapping varies with clone, age of the tree and frequency. Trees under high frequencies of tapping can also be converted to LFT. In high yielding clones by three annual stimulations (April/May, September and November) and medium yielding clones four annual stimulations (April/May, August, October and December) are recommended under third daily (d3) tapping frequency. Under d4 frequency, one panel can be tapped for at least 8 years and under d6 frequency the duration can be increased further to 10 years. The success of LFT depends on regular tapping throughout the year under the selected frequency and timely fixing of good quality rain guarding.

11.17.2 Intensive Tapping

Intensive tapping is normally carried out in old rubber trees for last few years before elimination. The intensive tapping method to be adopted generally depends on the availability of the bark in the tree, condition of the tree and previously adopted tapping system. The procedures adopted for intensive tapping are increased tapping frequency and expansion of tapping cut. The tapping system is further intensified by opening of double cuts and by using yield stimulants. At least 45 cm spacing is maintained between two cuts for proper drainage.

11.17.3 High Level Tapping

After continuous tapping the basal panel becomes uneconomic. So, the new cuts are made at higher level by using small ladders. In case of budded plant, the height should be 180 cm. When tapping of renewed bark on basal panels becomes uneconomic, new cuts are opened at higher levels, 180 cm from bud union or even higher. The tapper uses a small ladder to reach the cut. A tapper is usually able to tap only 135 trees in a day by using ladder.

11.17.4 Controlled Upward Tapping (CUT)

It is one of the modified tapping systems in which around 50% increase in yield can be achieved in long term. This method is adopted in plantations having low yield from the renewed bark and the bark is unsuitable for tapping. The bark consumption is minimum and higher yield can be obtained for longer period.

A long handled modified gouge knife is used to make the cut. The length of the tapping cut would depend on the duration for which controlled upward tapping is to be done. There are lots of realistic troubles associated with 1/2 spiral cut and thus it is better to tap quarter or one-third spiral cuts. As it is not easy to rain guard the CUT panel, it is advised to give rest the panel during rainy season. Tapping is done on the upper cut with maximum control on bark consumption (not to exceed 3 cm/month) and maintenance of the angle of cut at 45°. The tapping cut is opened on the virgin bark just above the renewed bark of the base panel. The tapping frequency is the same as that of base panel, i.e., d3 frequency for high yielding clones and d2 frequency for medium and low yielding clones. Tapping cut in the high panel can be encouraged using 5% ethephon following lace application method.

11.17.5 Rain Guarding

In rainy season tapping process needs to be carried out by fitting polythene rain guard to the trunk just above the tapping cut. By fixing an appropriate channel on the trunk just on top of the tapping cut, flow of water all the way through the main trunk can be checked out. By adopting this technique, the tapping cut and the bark below are kept in dry condition during the rainy season. The frequency of tapping and its effectiveness also depends on timely rain guarding. The rain guard material must be tested for phytotoxicity. Ready to use quality rain guard compound should be used. There are four types of rain guards, i.e., polythene skirt, tapping shade, guardian rain guard and tapping shield that are recommended for the rubber plantation.

11.18 Processing

Tapping leads to flow of latex from the rubber plant and the latex is collected in the coconut shell cup which is attached to the plant. Initially the latex is slightly alkaline or neutral but suddenly it becomes acidic in nature quickly due to the bacterial action. The fresh latex cannot be stored as such for long time without coagulation. As per requirement anticoagulant like ammonia, sodium bisulphate and formalin are added. Ammonia is recommended for preserved latex or latex concentrates while for rubber sheet processing sodium bisulphate is recommended. Later, the collected latex is transferred to clean buckets after two to three hours of tapping. The dried latex that is found on tapping panel is also collected by the tapper before tapping. The dried-up latex that are found on the ground (earth scrap) due to overflows is also collected in one month interval. The tree lace, shell scrap and earth scrap together account to 10-20 %. The collected latex is then centrifuged at 1440 rpm by which the liquid portion comes out. The concentrated latex is collected and chemical is mixed to enhance the storage life. The concentrated latex consists of 60% latex. The skim latex is collected in a different tank and then the sulphuric acid is added to get skim latex. Skim latex consists of 4-6% dry rubber and fetches low price in the market due to its poor quality whereas concentrated latex fetches higher price. Latex is collected from the cup and processed into crepe rubber, sheet rubber, processed latex, latex concentrate or block rubber as per requirement.

A. Sheet rubber

The collected latex is poured into coagulating tanks or small aluminium acidic acid or formic acid is used for coagulation. In order to prevent the surface darkening, sodium bisulphate 1.2g/kg is added to the diluted latex. Slow coagulation makes soft rubber, which helps the workers to work on the rollers. After coagulation, the rubber sheets are frequently washed down a number of times with alterations of water and later it is passed through hand or power operated rollers. The excess water and other dissolved impurities are pressed out from the sheets. Then the sheets are treated with paramitro phenol (0.5-1%) to prevent the mould growth. The surface of the sheets may be either smooth or grooved or zig zag or straight depends on the roller. These sheets are spread out in shade for two to three hours. The rubber sheets are then carried to smoke houses for proper drying and to avoid development of blisters. These are known as smoked sheets or dry ribbed sheet rubber. The process is carried out for 4-6 days. The sheets are smoked at a low temperature of 48-50^{0}C in the first day and then the temperature is being maintained 68^{0}C with low relative humidity in 2nd to 4th day. The various grades of rubber sheets are RMA IX, RMA-1, RMA-2, RMA-3, RMA-4 and RMA-5. The high-grade rubber sheets are characterized by clear, free from blisters, translucent and of a golden brown colour and fetch a better price.

B. Crepe Rubber

The coagulated latex or field coagulum collected are allowed to pass through heavy rolls to get a lace like rubber. After air drying it is called as crepe rubber and graded depending upon the type of material used.

i. **Pre-Coagulated Crepe:** It is prepared from the field latex after proper sieving and later it is bleached with euro bleach. The undesired material is then removed by adding little acid and further it is coagulated by adding the desired quantity of acid. The coagulum is then passed through a creeping battery that should consist of at least one macerator, one intermediate crepe rollers and finally one smooth roller to achieve a thin crepe. Later it is dried and packed.

ii. **Dry Crepe Rubber:** The coagulated latex or field coagulum is collected and then preliminary treatment is done by passing through a set of creping machines to get crinkly, lace-like rubber called crepe rubber. Different grades of crepe rubbers are EPC super 1 X, EPC 1X, EPC 2X and EPC 3X.

iii. **Sole Crepe:** The pre-coagulated crepe is cut as per required size and placed one above the other. They are then pressed with help of rollers. Then it is placed on laminated table which is heated rapidly by circulation of hot water. Later it is pressed with the help of rollers.

iv. **Pale Latex Crepe:** The fresh coagulation that is from diluted latex is treated with euro bleach and sodium bisulphate and then milled into thin crepe with the help of oil battery. Then this crepe is properly air dried and graded into 4 grades.

v. **Remitted Crepe:** The coagulum, unsmoked sheet and cup lumps are washed properly and milled in creping battery. The crepe obtained is air-dried, graded into 3 grades and packed.

vi. **Flat Bark Crepe:** All low-grade scrap and earth scrap are processed. It is washed properly to remove the dirt and then passed through crepe battery. Later the crepe is air dried, graded and packed.

vii. **Estate Brown Crepe (EBC):** The cup lumps and higher-grade field coagulum are collected and soaked for 24 hours. It was washed properly to remove the dirt and then passed through crepe battery. Later the crepe is air dried, graded and packed.

C. Latex Concentrates

The fresh latex that is collected immediately from the rubber plantation contains only 30 to 40 % dry rubber content and the rest is water. Due to high percentage water content, it becomes a delicate material and is easily damaged by bacterial infection. The concentration is developed in order to preserve the material for longer period. Different equipment and methods are employed for creating suitable latex concentrates with a dry rubber content up to 60%. Some preservatives are also added to increase the shelf life of the produce.

D. Creamed Concentrates

This can be achieved by mixing with ammonium alginate or tamarind seed powder solution is added to ammonia preserved field latex. It is a slow method but easily adopted in small scale rubber production centres. Creamed concentrates contain 50-55% drc and are suitable for thread rubber production.

E. Centrifuged Latex Concentrates

In this method the preserved latex is separated into two fractions, using a high-speed centrifuge, one containing the concentrated latex (60% drc) and the other containing serum (4-8% drc). These types account for more than 90% of world latex concentrates production.

References

Abraham J, Joseph K and Joseph P 2015. Effect of integrated nutrient management on soil quality and growth of *Hevea brasiliensis* during the immature phase. *Rubber Science,* **28** (2):159-167.

Aibara I and Miwa K 2014. Strategies for optimization of mineral nutrient transport in plants: multilevel regulation of nutrient-dependent dynamics of root architecture and transporter activity. *Plant Cell Physiol.,* **55** (12): 2027-2036.

Bissonnette J-F and De Koninck R 2017. The return of the plantation? Historical and contemporary trends in the relation between plantations and small holdings in South-East Asia. *J Peasant Stud,* **44** (4): 918-938.

Boedt L 2001. *Hevea brasiliensis* Muell. Arg. In: RH Raemaekers, ed.: *Crop Production in Tropical Africa*, Directorate General for International Cooperation (DGIC), Brussels, 1106-1131.

Carr MKV 2011. The water relations of rubber (*Hevea brasiliensis*): a review. *Exp Agric,* **48** (2):176 -193.

Chandrasekera LB 1984. Intercropping Hevea replantings during the immature period. *Proceedings of the International Rubber Conference*, Colombo, **1**(2): 389-393.

Debasis M, Datta B, Chaudhury M and Dey SK 2015. Nutrient requirement for natural rubber. *Better Crops,* **99** (2):19-20.

Edathil TT, Jacob CK and Joseph A 2000. Leaf diseases. In: *Natural Rubber: Agro Management and Crop Processing* (Eds. PJ George and C Kuruvella Jacob). Rubber Research Institute of India, Kottayam, pp 273-296.

Faiz MAA 2006. Efficacy of glyphosate and its mixtures against weeds under young rubber forest plantation. *J. Rubber Res.,* **9**: 50–60.

Gururaja Rao G, Sanjeeva Rao P, Devakumar AS, Vijayakumar KR and Sethuraj MR 1990. Influence of soil, plant and meteorological factors on water relations and yield of *Hevea brasiliensis*. *International Journal of Biometeorology,* **34**:175-180.

Jessy MD, Punnoose KI and Nayar TVR 2001. Crop diversification and its sustainability in young rubber plantations. *Journal of Plantation Crops*, **33** (1): 29-35.

Manju MJ, Sadananda M, Santhosh HM, Patil RS, Shankarappa TH, Benagi VI and Idicula SP 2019. Evaluation of different fungi toxicants against *Corynespora cassiicola* causing Corynespora Leaf Fall (CLF) disease of rubber. *Int. J. Curr. Microbiol. App. Sci.*, **8** (02): 1640-1647.

Meti S, Gohain T and Chaudhuri D 2002. Response of *Hevea* to fertilizers in northern West Bengal. *Indian J Nat Rubber Res,* **15** (2):119-128.

Purseglove JW 1977. *Tropical Crops: Dicotyledons*. Longman Group, London, Third Edition, London, 719p.

Pushparaja E 1983. Problems and potential for establishing *Hevea* under difficult environmental conditions. *Planter*, **59**: 242-251.

Rajasekharan P and Veeraputhran S 2002. Adoption of intercropping in rubber small holdings in Kerala, India: A to bit analysis. *Agroforestry Systems*, **56** (1): 1-11

Razman MZ, Wan Daud WMN and Sulaiman Z 2016. Effect of mulching and fertilizer rates on the growth of RRIM 3001. *Int J Agric For Plant,* **4**: 33-37.

Rodrigo VHL, Silva TUK and Munusinghe ES 2004. Improving the spatial arrangement of planting rubber (*Hevea brasiliensis* Muell Arg.) for long-term intercropping. *Field Crops Research*, **89**: 327-335.

Viswanathan PK 2008. Emerging small holder rubber farming systems in India and Thailand: A comparative economic analysis. *Asian Journal of Agriculture and Development,* **5** (2):1-16.

Vimalakumari TG 2001. Influence of intercropping on the rhizosphere microflora of *Hevea. Indian Journal of Natural Rubber Research*, **14** (1): 55-59.

Venkatachalam P, Jayashree R, Rekha K, Sushmakumari S, Sobha S, Yayasree PK, Kala RG and Thulaseedharand A 2007. *Methods in Molecular Biology: Rubber Tree* (*Hevea brasiliensis* Muell Arg.). Humana Press (now Springer Science, New York), pp. 153-164.

Verheye W 1999. *Environmental Study of Palmci Plantations in Ivory Coast*. Mission Report, Geography Dept., University Gent, Belgium, 54p.

Yew FK 1982. *Contribution towards the Development of a Land Evaluation System for Hevea brasiliensis* Muell Arg. *Cultivation in Peninsular Malaysia*. PhD thesis, University Gent, Belgium, 261p.

Webster CC and Paardekooper EC 1989. The botany of the rubber tree. In: Rubber, 57–84 (Eds C C Webster and WJ Baulkwill). Essex, UK: Longman

OUTCOME ASSESSMENTS

PART A

Answer the following questions true or false.

1. In rubber plantations major deficiency symptoms are seen due lack of copper. True/False
2. Intensive tapping is normally carried out in old rubber trees. True/False
3. Latex is found in the phloem in the form of small particles. True/False
4. Rubber performs best under adequate moisture condition. True/False
5. Amazon basin in Brazil is the centre of origin of rubber. True/False

PART B

Answer the following questions.

1. What is the causal organism of powdery mildew in rubber?
2. In rainy season tapping process needs to be carried out by fitting — to the trunk just above the tapping cut.
3. What does d2 mean in tapping?